25
4.9

Answering Questions in Physical Chemistry

Answering Questions in Physical Chemistry

F. H. GOODYEAR
B.Sc., A.R.I.C.
*Lecturer in Physical Chemistry,
North Staffordshire College of Technology,
Stoke-on-Trent*

Heinemann Educational Books Ltd: London

Heinemann Educational Books Ltd
LONDON EDINBURGH MELBOURNE TORONTO
AUCKLAND SINGAPORE JOHANNESBURG
HONG KONG IBADAN NAIROBI

SBN 435 66370 4

© F. H. Goodyear 1969
First published 1969

Published by Heinemann Educational Books Ltd
48 Charles Street, London W1X 8AH
Printed in Great Britain
by Richard Clay (The Chaucer Press), Ltd
Bungay, Suffolk

Preface

THE book is planned to help all students working for examinations in physical chemistry at what might be conveniently called 'Part I level'. This includes Part I examinations of Special or Honours degree courses and of the Graduateship examinations of the Royal Institute of Chemistry, Part II examinations of General degree courses, Higher National Certificate and Higher National Diploma final examinations, and appropriate parts of technological degree courses approved by the National Council for Academic Awards.

The author gratefully acknowledges the courtesy of the Council of the Royal Institute of Chemistry and of the Senate of the University of London in granting permission for questions from Grad. R.I.C. and various degree examination papers to be reproduced in this book. Neither the University nor the Royal Institute is in any way committed to approval of the answers or notes concerning any of their questions.

The source of each question is indicated in accordance with the following scheme:

- C   Higher National Certificate in Chemistry
- D   Higher National Diploma in Chemistry
- Ge London University, B.Sc. General, Part II (External)
- Gi London University, B.Sc. General, Part II (Internal)
- S   London University, B.Sc. Special, Part I (External)
- R   The Royal Institute of Chemistry, Grad. R.I.C., Part I.

The author also wishes to thank his colleague, Dr. J. G. Dawber, for reading through the manuscript and making some useful suggestions.

Contents

Introduction	ix
Chapter 1: The Theory and Technique of Examinations	1
1.1 The Time Factor	1
1.2 The Accumulation of Marks	2
1.3 Types of Question and Style of Answer	5
1.4 Choice of Questions	7
1.5 Choice of Examples	9
1.6 Diagrams and Graphs	9
1.7 Language and Style	12
1.8 General	13
Chapter 2: Questions in Physical Chemistry	15
2.E Electrochemistry	15
2.K Chemical Equilibrium and Kinetics	32
2.M Molecular and Crystal Structure	42
2.P Phase Rule	45
2.S Surface Chemistry	53
2.Sp Spectra	59
2.T Thermodynamics	63
2.W Wave Mechanics and Quantum Theory	76
2.X Miscellaneous	79
Chapter 3: Annotated Full Answers to Selected Questions	86
Chapter 4: Sign and other Conventions	102
Appendix: Recommended Symbols	105
Index	109

Introduction

THERE are many good books of calculations in physical chemistry, some of which also include useful notes on the relevant theory, but hitherto there appears to have been no published attempt to give guidance to students in answering non-numerical questions. That such guidance is necessary is shown by the comments of examiners. For example, the Examiners of the Royal Institute of Chemistry have remarked on 'the poor structural form for essays' and that 'instructors must give more guidance and preferably practice in essay writing; the present standard is simply not good enough'. Furthermore, 'candidates have completely misjudged the standard required'. There are also complaints of irrelevance in answers and of candidates not answering the questions asked, or, at least, only in part.

Such comments are, unfortunately, not isolated remarks, but recur regularly. One examiner deplores the 'dangerous trend' shown by 'the sacrifice of a knowledge of chemical behaviour to the acquisition of theoretical knowledge', and adds that there is 'no point in knowing the principles of chemistry if you know no chemistry'. This is perfectly true; one might add that there is no point in knowing the principles or the chemistry if you are unable to express yourself clearly in good English. Lack of communication is as disastrous in chemistry as in international affairs.

No one who has taught in a College of Technology will disagree with the Examiners of the Royal Institute, but very few indeed take steps to remedy the situation. It is hoped that this book will encourage them to do so.

Moreover, many lecturers make considerable use of books of calculations, particularly in setting homework, it being generally believed that physical chemistry theory becomes understood only when it is applied. Laboratory work is also recognized as partly designed to test the student's application of previously explained theory. It is now suggested that considerable benefit may be obtained by answering non-numerical questions also.

A survey of examination papers over the past five or six years shows the following percentages of questions containing numerical problems:

London University
  B.Sc. General, Part II (External)   4·5 per cent
  B.Sc. General, Part II (Internal)   7·0 per cent
  B.Sc. Special, Part I (External)   4·3 per cent

Royal Institute of Chemistry
  Grad. R.I.C., Part I   21·5 per cent
  Grad. R.I.C., Part II   32·0 per cent

These figures reveal interesting differences. Calculations are much more popular with Royal Institute of Chemistry examiners than with London University examiners. In fact, almost half the papers set by the latter contain no calculations whatever.

Books of calculations in physical chemistry give a false impression of the kind of questions likely to be met in actual examinations. In the first place, they encourage a student to believe that calculations form the main part of the examination, a fact which the above figures disprove. Secondly, they cannot show the relationship between the numerical part of a question and the rest of the question, where, as is usually the case, the calculation does not form the whole of the question. This relationship is often very important. The two parts are interdependent, either or both often giving a clue as to how the other section may be tackled. Some examples of this will be referred to as they arise in Chapter 2.

It is, of course, pointless to repeat calculations which need only the insertion of numerical values in a memorized formula. Likewise, there is little use in asking homework questions which merely require the reproduction of lecture notes. Any competent examiner knows how to set questions which involve much more than such reduplication, which require the assembly of facts from different sources and give a candidate the opportunity of introducing material garnered from his much wider reading of the subject. It is hoped that the questions assembled in Chapter 2 will be found to be mainly of this kind and eminently suitable for the training of physical chemists.

Arrangement of this Book

Very little has been written on the theory of examinations, or, indeed, on the techniques. Some brief comments on these subjects are made in Chapter 1.

The arrangement of the questions in Chapter 2 calls for some explanation. They have been divided between nine topics labelled as far as possible with the initial letter of the name of the topic (for example, 'E' for Electrochemistry).

In view of what has been said above concerning the nature of examination questions, it will be evident that many answers will require facts from more than one of the 'topics'. Also, some questions could fit equally well into either of two (or more) sections. Such questions have been included under the most convenient topic and cross-referenced at the end of the questions under the other topics concerned.

Each of the sectional topics in Chapter 2 is sub-divided in the same way. The first two sub-sections are lists of (i) definitions and (ii) laws and equations, with occasional remarks as necessary. The purpose of these lists is twofold: they are a reminder of the facts that may be needed in

INTRODUCTION

answering the questions that follow but they should also form a useful compilation for revision purposes.

The lists may be used in various ways; for example, a student may wish, during his final revision period, to attempt to write out on a sheet of paper the definitions, laws, and equations listed. Alternatively, he may prefer to note by the side of each item a few essential words of a definition or law, or an actual equation. Complete definitions or laws should *not* be inserted; if this had been considered desirable they would have been printed in full. The present purpose is not to replace a textbook or lecture notes but to encourage students by suggesting what is important to learn, and by occasionally jogging the memory.

The actual examination questions then follow as the third sub-section. For many of the questions notes are provided on difficult or doubtful points, but it must be emphasized that there are always several ways of answering a question, and there may be alternative arrangements of the facts, or even different facts in some cases, which would be equally acceptable.

In Chapter 3 are given complete answers to a few selected basic questions. It is not claimed that these are perfect answers, if, indeed, such things exist. The purpose of this chapter is to illustrate the points made in Chapter 1 by means of suitable examples. The questions have been chosen from different topics, and answered in appropriate styles. The appended notes are intended to:

(*a*) draw attention to the application of a principle mentioned in Chapter 1;
(*b*) suggest alternative treatments or possible extensions; or
(*c*) explain technical matters.

Whenever notes or full answers are given the student is advised to prepare his own outline or full answer before referring to the notes or to Chapter 3.

It is suggested that a student should generally be required to answer questions at a length appropriate to the examination for which he is studying, but occasionally it might be a useful exercise to write a much longer answer on one of the more important topics. Such answers should be based on wide reading around the subject. A certain amount of literature research of this kind results not only in a better understanding but also in a clearer recollection of details at examination times.

Questions have been reproduced as nearly as possible as published in the original question papers, *including any errors, ambiguities, or inconsistencies*. The notes following the questions will draw attention to any errors and ambiguities. Inconsistencies rarely occur within one question, but all too frequently between questions of the same examining body.

The more obvious inconsistencies are in the use of abbreviations and capitals, and in such matters as the methods of writing cell diagrams in electrochemistry. The objection is that it sets a bad example. Students should aim at complete consistency in these things. The use of abbreviations will be referred to in Chapter 1. Capitals should be used sparingly.

The book concludes with an explanation of sign conventions (Chapter 4) and a reference appendix of recommended symbols.

1

The Theory and Technique of Examinations

Some people may not realize that there is anything worth learning about the theory and techniques of examinations, and certainly teachers at all levels often fail to give instruction in these matters. This no doubt arises mainly from overloaded syllabuses, but much valuable guidance could be given to students even in a one-hour lecture. Whatever may be the reason for the omission, this chapter is designed to rectify it.

1.1 The Time Factor

Every examinee will appreciate the importance of the time factor, but it is surprising to see how often it is overlooked or ignored. The time allowed to complete an examination paper is, of course, strictly limited and there is nothing that a candidate can do about that. But within this limit there are certain rules which should be followed in order to make the most of the time available.

In examinations of the standard with which we are now concerned it is customary for candidates to be asked to answer four or five questions in three hours, out of a total of eight or nine questions on the paper.

The first task must be to read right through the paper, for which an allowance of at least five minutes should be made. If five questions are to be answered this will still leave thirty-five minutes for each question. It cannot be too strongly emphasized that the full quota of questions *must* be attempted. On analysis of the figures for one public examination selected at random it was found that between eight and nine per cent of candidates did not answer the required number of questions. As in this case only four answers were asked for, these candidates had themselves reduced their maximum possible marks to seventy-five per cent. But, of course, full marks are rarely given for an answer, and a grave risk is taken by forgoing the opportunity of gaining at least a few of those twenty-five forfeited marks.

Failure to answer the full quota of questions may sometimes be due to lack of knowledge or inadequate preparation for the examination, but sometimes it is a matter of bad timing.

One must point out, however, that the statistics referred to above do not reveal how many attempts at questions were merely a hastily scribbled line or two in the last few minutes of the examination. Regrettably, this is an all-too-common sight for examiners, and one which candidates are advised to avoid.

1.2 The Accumulation of Marks

An examiner usually works to a marking scheme in which the various sections of an expected answer are allocated definite maximum marks. For example, an essay may be awarded a maximum of seventeen marks for the subject matter and three for presentation. The seventeen marks may be subdivided to cover the important facts which are considered to be essential. A bonus mark may be given for additional information, especially if it indicates wide reading of the subject. Some examiners also deduct marks for omissions or inaccurate statements.

For a question such as T.12 (selected at random from Chapter 2) the marking scheme might be:

Explanation of chemical equilibrium	3 marks
Explanation of equilibrium constant	3 marks
Derivation of equation	7 marks
Method of calculation	2 marks
Result of calculation	5 marks
Total	20 marks

Other examiners might hold different views as to the relative importance of the sections of this question, perhaps giving more marks for the calculation and fewer for deducing the equation, or not making any allowance for the correct method of calculation. Unfortunately, examiners do not disclose their marking schemes, but it is clear that candidates must check their answers in order to be sure that no part of the question has been overlooked.

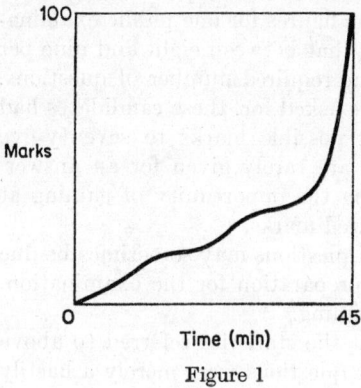

Figure 1

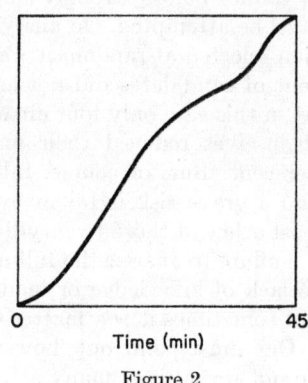

Figure 2

THE THEORY AND TECHNIQUE OF EXAMINATIONS

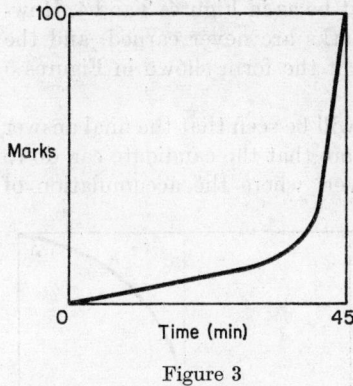

Figure 3

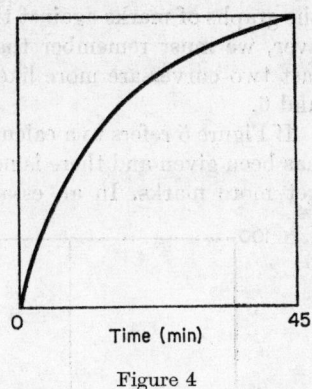

Figure 4

Whatever the marking scheme may be we can expect that, in general, marks will accumulate according to a certain pattern, depending on the type of question.

Let us consider some examples. For a calculation in which the final figure is the most important part of the answer, we might expect the graph of marks against time to be of the form shown in Figure 1. Marks are being steadily accumulated by following the correct procedure, step by step, with mathematical accuracy, but the last step brings in the most marks, so that the curve rises rapidly near the end. This, of course, is the ideal curve for the perfect answer gaining full marks. If a serious mistake is made half-way through the calculation the final mark might be very low indeed.

In contrast to the calculation we might look at an essay question. Here a few marks would be earned for the introduction, most marks for the body of the essay, and possibly a few for the conclusions. The resulting graph of marks against time would be as shown in Figure 2.

We might now suppose that in the two hypothetical extreme cases

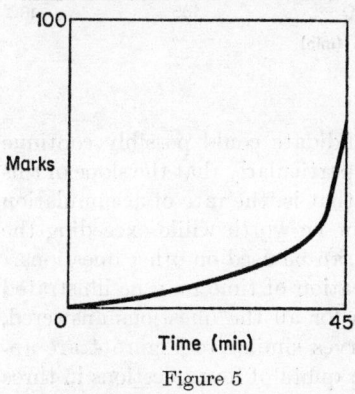

Figure 5

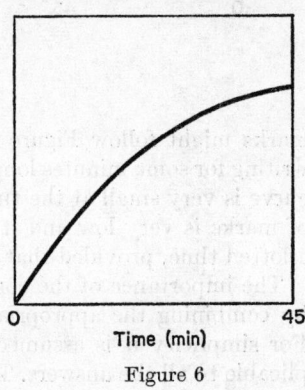

Figure 6

the graphs of marks against time might be as in Figures 3 and 4. However, we must remember that full marks are never earned, and the last two curves are more likely to be of the form shown in Figures 5 and 6.

If Figure 5 refers to a calculation it will be seen that the final answer has been given and there is nothing more that the candidate can do to get more marks. In an essay, however, where the accumulation of

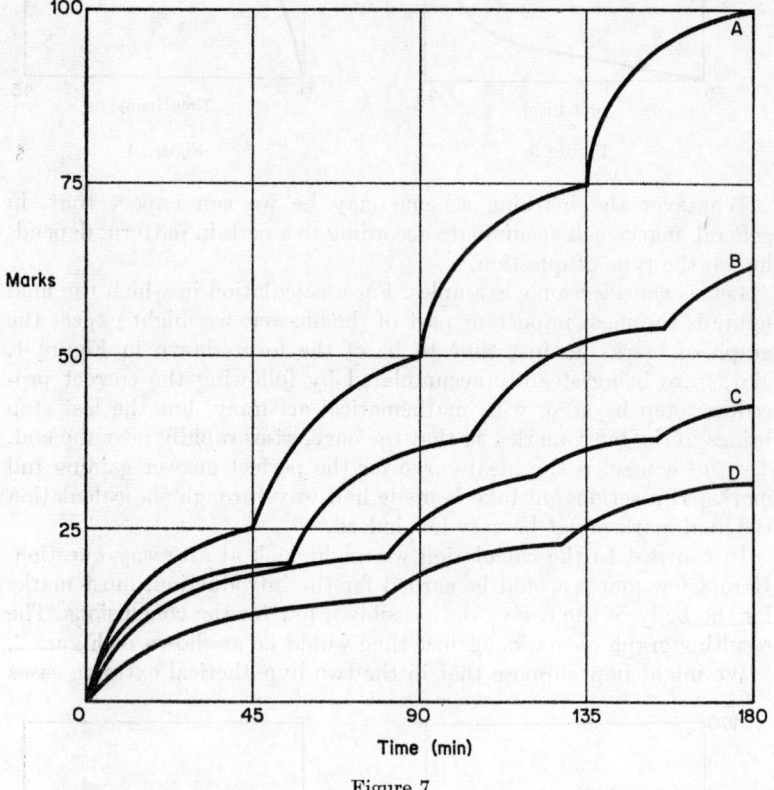

Figure 7

marks might follow Figure 6, the candidate could possibly continue writing for some minutes longer. Note particularly that the slope of this curve is very small at the time limit, that is, the rate of accumulation of marks is very low and it would not be worth while exceeding the allotted time, provided that the time can be used on other questions.

The importance of the correct allocation of time may be illustrated by combining the appropriate curves for all the questions answered. For simplicity it is assumed that curves similar to Figure 4 are applicable to all the answers. Then, for a quota of four questions in three hours Figure 7A shows the ideal result. A more likely result for an average student would probably be represented by curve 7B. If far too much time was spent on the first question the result might be curve 7C, even though good answers are given, as far as time permits for the fourth question. It is easy to see in curve 7D how disastrous poor answers *and* poor timing can be.

1.3 Types of Question and Style of Answer

Practice in answering non-numerical questions is just as important to examination candidates as is practice with calculations. Much time is wasted in examinations simply because a candidate is not familiar with the kind of question asked, or, more important, with the kind of answer required. In practical classes students are shown how to write reports of the work done, even though such reports are often limited by the familiar headings 'Theory', 'Method', 'Results', 'Conclusions'. This at least has the advantage of saving time; the *form* of the report is predetermined.

In answering examination questions one has to decide on the form of the answer as quickly as possible so that the facts to be presented may be marshalled in an appropriate order. Reference has already been made to examiners' comments concerning the poor quality of essays, and similar observations could be made on other kinds of answers.

The essay is a literary form which has a standard structure, though probably not as rigid as, for example, the sonnet or the limerick. It should open with one or two introductory sentences, the theme should be developed in the body of the essay, and the whole rounded off by a short paragraph of conclusions or summary. The sections should not be headed; in fact, there are two schools of thought concerning headings in essays of a technical nature. There are those who believe that a few headings may be necessary to a clear presentation of the subject, whereas others consider that if an essay is intended it should follow the recognized form and be entirely free from headings and sub-headings. Not knowing to which school the examiner might belong, it is advisable for examination candidates to avoid using headings in an essay.

Questions which do not directly ask for an essay might, nevertheless, be conveniently answered in essay form. One might, for example, be asked to 'review' or 'write a critical survey of' a subject, and it is left to the candidate to decide whether to write a formal essay or an informal discourse making frequent use of headings and sub-headings. The latter might be the quicker, especially where, for example, a review of experimental methods is required.

At the other extreme is the question asking for 'notes' or 'informative notes', whatever the difference might be. This does not mean 'reproduce your lecture notes on this subject', but that one should show the extent of one's knowledge of the subject by brief statements; that is, brief but grammatical.

A 'discussion' may be of any literary form but must look at the subject from all sides. In chemistry this will not often mean from different points of view, but rather in a critical manner. For example, the question might be 'Discuss the methods available for the determination of the pH of a solution.' One could list the methods in five minutes, but this would be of little value. The discussion might involve explanations of the theoretical bases for the methods, the sources of errors and their avoidance, and the estimation and comparison of accuracy, sensitivity, and reproducibility.

The derivation of an equation is probably the only occasion on which the reproduction of lecture notes or textbook material is permissible. A derivation must follow a perfectly logical course and generally there can be no deviation from this course, although there may sometimes be alternative derivations, perhaps from different starting points.

Laws and definitions are frequently asked for as part of a question, and these should be known by candidates so that no time is wasted in 'scratching' for them. The important laws, equations, and definitions have been listed in each section of Chapter 2 in order to form a ready reference when answering the questions, and also to act as a useful revision list. It should be emphasized, also, that probably the majority of answers should begin with a definition. An answer to the question referred to above could well begin by defining pH. This will show its relationship to hydrogen ion concentration and will lead directly to some of the methods for its measurement.

Giving a definition at the start of an answer focuses the writer's attention and often brings to mind important related facts which can usefully be incorporated in the answer. It is a practice that is strongly recommended whenever appropriate.

According to the *Oxford Dictionary* a definition is 'a statement of the precise meaning of a term'. Precision is certainly necessary, but there are other essential criteria of a good definition. We might state these as follows:

A good definition must be:

   (*a*) concise, and therefore easily memorized and reproduced, and containing no unnecessary words;
   (*b*) precise;
   (*c*) complete; it must contain all the necessary and sufficient conditions.

Consider an example: 'The order of a reaction is the number of species whose concentrations determine the rate of reaction.' It we break down this definition it will be seen that each word is essential and that the statement is complete; no further qualification or modification is necessary. In fact the whole statement is summarized by the four words—'number', 'species', 'concentrations', 'rate'—conveniently for revision purposes.

Compare the above definition with an alternative: 'The order of a reaction is the number of different kinds of atoms, molecules, or ions whose concentrations have to be inserted in the rate equation for the reaction concerned.'

Is this concise? Obviously not, since it contains almost twice as many words as the first definition. It is not necessary to use the words 'different kinds of' nor 'for the reaction concerned', and, as we have seen, the word 'species' is much more concise than 'atoms, molecules, or ions'. Finally, when the above words are omitted the definition is not precise and just misses the mark. The concentration terms which need to be inserted in the rate equation are just those of the species which determine the rate of the reaction. So we conclude that, while the second definition is not *incorrect*, it is not a good definition.

1.4 Choice of Questions

Candidates will naturally choose the questions which they feel they can answer best, and each will have his own particular preferences as far as subjects go. Hence very little advice can be given in this connection.

The only question which should be completely dismissed at first reading of the paper is that which has not been covered in the candidate's course of study. There is a possibility of this occurring in external examinations. Subsequent sorting of questions depends on the candidate's personal preferences, but there are one or two points worth consideration.

The very wide, generalized question is often the most difficult to answer well. It has been known for a question at Higher National Certificate or Higher National Diploma level to be set on the lines of: 'Describe in detail any one experiment which you have carried out in the physical chemistry laboratory.' Such a question is a snare and a delusion. With such a free choice of subject it is clear that a very high standard of marking will be applied, and the selection of the experiment will be open to severe criticism. Unless one has successfully carried out a small 'project' type experiment, out of the usual run of things, and with more than usually interesting results, one would be well advised to ignore this question. Such questions are not (or should not be) designed as free gifts to the weaker candidates, but opportunities for better candidates to shine.

The Part I examinations of most examining bodies have two main purposes. In the first place they are designed to eliminate those who are not likely to benefit from the further one year of full-time, or two years of part-time, study for the Part II examinations, but they are also intended to test the basic knowledge of a wide field of chemistry which is needed before delving more deeply into some of the topics. For these reasons a wide choice of questions is given and inevitably questions on the more important topics are bound to appear regularly.

An analysis of questions appearing over the last few years, in the four series of papers from which most of the questions in Chapter 2 are taken, gives the results shown in Table 1.

TABLE 1

TOPICS IN EXAMINATION QUESTIONS

Average number of questions per paper in the topics* :

Exam.	E	T	Sp	K	W	S	M	P
Ge	1·8	1·6	0·7	1·2	1·0	1·5	0·4	0·4
Gi	2·0	1·6	0·9	1·1	0·6	1·0	1·0	0·2
S	1·4	2·3	1·4	1·0	1·2	0·4	0·4	0·4
R	2·5	2·2	(0·2)	1·1	0·1	1·3	0·1	1·6

* *See* code listed in the Contents, page vii.

In the Royal Institute of Chemistry Part I examinations questions on spectra have been introduced only from 1966, so that the figure given for these in Table 1 is not significant.

The analysis shows a general pattern; for example, 'Ge' and 'Gi' papers (B.Sc. General) are very similarly constituted, with the emphasis on electrochemistry and thermodynamics, but 'S' papers (B.Sc. Special) show a stronger bias towards thermodynamics, spectra, and wave mechanics (including quantum theory, etc.). On the other hand, 'R' papers (Royal Institute), while still emphasizing electrochemistry and thermodynamics, contain very little on wave mechanics or molecular structure, preferring the phase rule, surface chemistry, and kinetics.

While these statistics are of great interest to most examination candidates, a word of warning must be given. It would be very foolish to rely too much on the continuation of the general patterns of recent years. A change of examiners might, without warning, produce papers with an entirely different emphasis, although most examining bodies have procedures for maintaining the general standard of papers and ensuring a fair spread of questions over the syllabus.

Not only do examiners have their own preferred subjects, but rapid progress in a particular branch of physical chemistry naturally results in a spate of questions on the basic theory.

Furthermore, questions on the five most popular topics frequently require a knowledge of other branches of physical chemistry. This theme of the interdependence of topics, and the danger of trying to keep each one in a watertight compartment, will recur repeatedly, and the student himself will realize its implications as he works through the questions in Chapter 2. In addition to this, it will be seen that many questions are in two or more parts on widely separated subjects.

A more fundamental reason for not neglecting the less frequently appearing topics is to be found in the second purpose of Part I examinations, namely to test the basic knowledge of chemistry. Students have been known to pass a Part I examination after revising only three or four of the main topics in physical chemistry, but these students invariably have to struggle throughout the Part II course, and usually fail the final examination.

1.5 Choice of Examples

Examples should be given whenever appropriate, but they should be carefully chosen. Generally, examples will have been noted when studying a subject. These should be memorized during revision so that they readily come to mind when the particular topic is mentioned, and it should rarely be necessary to spend time trying to think up examples.

Whenever possible an example should be clear-cut, not requiring any qualification nor introducing any extraneous ideas. If asked for an example of a first-order reaction choose a simple elementary reaction such as the decomposition of ethylamine, not, say, a reversible reaction such as the mutarotation of glucose. The latter might entail further explanations which are not really relevant to the question.

1.6 Diagrams and Graphs

Poor diagrams may be worse than useless, but a clear, reasonably proportioned diagram, suitably labelled, can be extremely useful and time-saving.

Artistic merit is not looked for in a diagram, nor is a scale drawing required, but so many efforts are far from artistic and just as far out of scale. It should be well within the powers of any budding chemist to draw something resembling his simple apparatus, of reasonable size and approximately to scale, not suggesting, for example, titration of a bucketful of solution with a micro-burette, nor of a thimbleful with a 100-ml burette. Yet this sort of *art nouveau* so often appears in examination scripts.

A sense of scale is even more important when drawing graphs. Graphs are similar to statistics; they may be used to prove almost anything. There are those (not only students) who can, by suitable choice of scales, prove a linear relationship between practically any two variables, to their own apparent satisfaction. Certainly a graph should display the relationship to the best advantage, but not so as to make what should be a curve appear to be a straight line.

The aim is always to try to find a simple linear relationship, but this may not always be possible. When a straight line is to be expected, the scales of the two axes should be so selected that the slope of the line will be near to 45°. In this region measured slopes will be more accurate, as also will be any extrapolated or interpolated readings from the

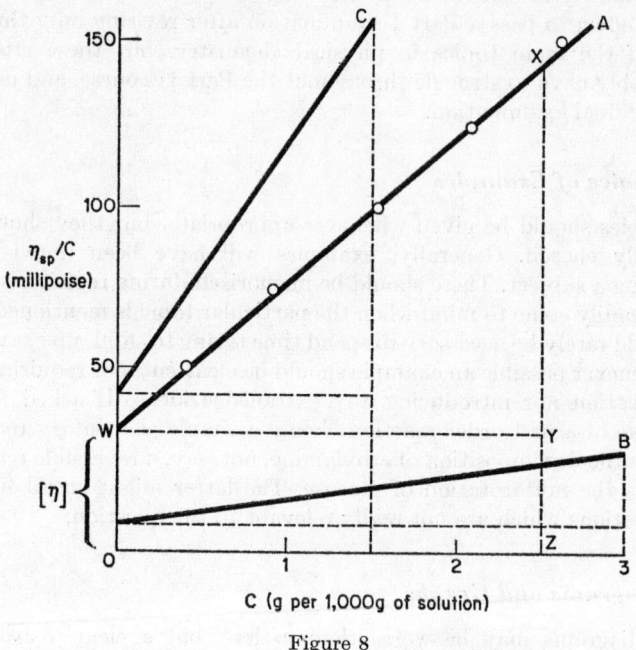

Figure 8

graph. With other kinds of curve it may be more difficult to choose the best scales immediately, but it is quite useful to aim at a slope of about 45° for a hypothetical straight line joining the first and last points of the graph.

In Figure 8, curve A shows the most convenient representation of the relationship between η_{sp}/c and concentration, where η_{sp} is the specific viscosity of a polymer solution in millipoises and c is the concentration (grammes of solute per 1,000 g of solution). The best straight line has been drawn for the experimental points shown. Published graphs such as this are necessarily very much reduced from the original size. In an examination paper the scales on the axes would probably need to be four or five times those of Figure 8. Note that the straight line has a slope near 45° which can be accurately measured by drawing (or imagining) the lines WY and XZ. The slope is, of course, XY/OZ. Curves B and C show how measurement of the slope becomes less accurate for very small or large slopes.

The intercept here is the intrinsic viscosity, $[\eta]$, and the equation to the curve is

$$\eta_{sp}/c = kc + [\eta]$$

which is of the form

$$y = mx + c$$

that is, a linear equation. Hence, one says that 'y is a linear function of x' or, here, η_{sp}/c is a linear function of c. Figure 9 also represents a linear function, that between the polarographic diffusion current and the concentration of a reducible substance. When this substance is absent the diffusion current is zero, i.e. the curve passes through the origin and $i_d = kc$, where k is the slope of the curve. Then we may also say that 'i_d is directly proportional to c'. It is important to distinguish between the relationships of Figures 8 and 9.

The best size for a graph is not easily judged by the inexperienced, but in examinations it is usually safest to aim at filling the sheet of graph paper as far as possible. But in doing so one must avoid selecting inconvenient scales. For example, when using graph paper of size 10 in. by 8 in., one might require to represent twenty-two units on the 8-in. axis. Dividing the number of units by the length might suggest a

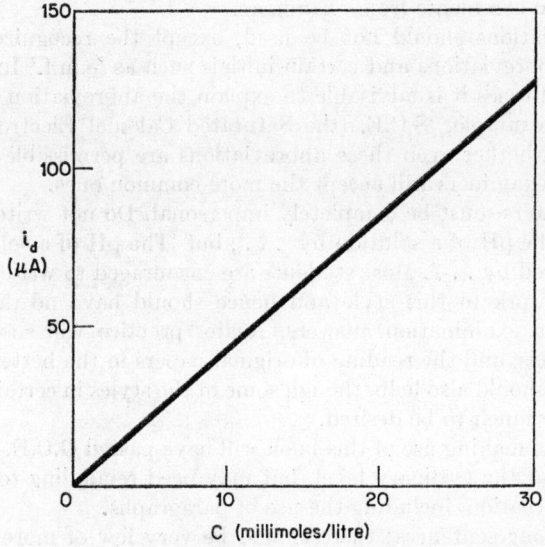

Figure 9

scale of three units to the inch. But this would be an extremely awkward scale to deal with, especially for interpolation. It would be far better to use four units to the inch, which would make use of nearly three-quarters of that axis. Alternatively, it might be better to turn the paper through 90° and use the 10-in. side for this axis.

The measurement of the slope of a straight line is a very simple process, but often very inaccurately done. The greatest accuracy will be attained by using the coordinates of two widely separated points, yet one regularly sees a small section in the middle of the curve used for this purpose. Often, also, thick lines are drawn, horizontally and vertically, to find the coordinates required, and these lines frequently deviate quite obviously from the printed lines on the graph paper. By choosing points as far apart as possible the relative error of the measurement is considerably reduced. The calculation of the slope can usually be simplified by using a whole number of units or squares on one axis and finding the corresponding number of units or squares on the other axis.

Finally, do not under any circumstances draw a graph without clearly labelling the axes, giving it a title or caption, and indicating the units used on each axis.

1.7 Language and Style

This book is not intended to be a thesis on English for chemists, but there are several matters which need to be mentioned as they frequently give rise to comments from examiners.

Abbreviations should not be used, except the recognized mathematical abbreviations and certain initials such as 'e.m.f.' In less well-recognized cases it is advisable to explain the abbreviation when first used, for example, 'S.C.E. (the Saturated Calomel Electrode)'. It is doubtful whether even these abbreviations are permissible in essays, but most examiners will accept the more common ones.

All answers must be completely impersonal. Do not write 'I should measure the pH of a solution by . . .', but 'The pH of a solution may be measured by . . .'. Most students are encouraged to write reports of practical work in this style and hence should have no difficulty in using it in examination answers. Again, practice will soon make a habit of this, and the reading of original papers in the better scientific literature should also help, though some of the styles in certain journals leave very much to be desired.

Students making use of this book will have passed G.C.E. in English language at the Ordinary level, but may need reminding to take care over punctuation, including the use of paragraphs.

Avoid long sentences; there should be very few of more than four lines in manuscript. Also avoid long paragraphs. Generally speaking a new idea should be introduced in a new paragraph, but a long discussion of one idea may need breaking down into shorter paragraphs. Remember that the end of a paragraph is a brief breathing space for the argument to be assimilated before proceeding further.

The repetition of words in a sentence or in successive sentences makes reading monotonous, and the repetition of statements or ideas, even in different wording, makes for even greater monotony. Further, any kind of 'padding' must be avoided. Do not be too concerned if an answer appears short when written; as long as it contains the necessary information concisely and clearly expressed it will be acceptable.

There are many words frequently occurring in physical chemistry which are commonly misused or abused. It is difficult to understand why the word 'statistics', for example, is so often incorrectly pronounced and spelt. Careless or incorrect pronunciation may be the cause *or* the effect of misspelling.

Some of the common words often incorrectly spelt are:

asymmetry	separate
desiccator	dependent
liquefy	

If in doubt try splitting the word into its component parts and sorting out its meaning. For example, 'asymmetry' is 'a-sym-metry', the parts meaning (roughly) 'not-the same-measure'. This clearly shows why there are two m's in the middle of the word and only one s.

Another very common fault both in speech and writing is the confusion of singular and plural forms of certain words of Greek or Latin origin. Unfortunately, these errors frequently occur in print and even find their way occasionally into learned journals. The correct forms of some such words are:

Singular	*Plural*
Spectrum	Spectra
Datum	Data
Bacterium	Bacteria
Criterion	Criteria
Phenomenon	Phenomena
Formula	Formulae

Do not use colloquialisms nor be facetious, and, finally, avoid dogmatism.

1.8 General

A too hasty reading of a question can lead to trouble of many kinds. Quite often a reasonable answer is given to a question which was not asked. The difference in wording may be quite small, but every word in the question is significant. Radiation chemistry, for example, is not photochemistry, though an essay on the former would include the latter. Make sure that the scope of the question is understood before starting the answer.

It is a good thing to refer repeatedly to the question while writing the answer. This helps to keep the answer on the right lines, and could reveal a fault which can be rectified before the question is finished with, so saving much time and many marks. It is quite easy, for example, to overlook a small section of a composite question.

The standard to be aimed at should be kept in mind, remembering that certain questions could be asked at any stage above G.C.E. level, but would require quite different answers according to the standard of the examination.

2

Questions in Physical Chemistry

Each topic in this chapter is sub-divided: (i) *definitions,* (ii) *laws and equations,* (iii) *questions,* (iv) *notes on the questions. Refer to the Introduction for a full explanation of this arrangement.*

E. ELECTROCHEMISTRY

ELECTROCHEMISTRY is a very wide topic which may be conveniently divided into three sections:

(a) The effects of the application of a potential difference or electric current to an electrolyte solution, i.e. conductance, electrolysis, electrodeposition.

(b) The production of an e.m.f. by chemical means (the reverse of (a)), i.e. galvanic cells, electrode potentials, etc.

(c) Equilibria in electrolyte solutions (in the absence of any electrical circuit or e.m.f.), e.g. ionic theory, electrolytic dissociation, hydrolysis of salts, pH, buffer solutions, indicator theory.

Division in this way is merely for convenience in studying the subject; the sections will clearly overlap. We shall generally keep to the arrangement without actually sub-dividing in the following paragraphs:

(i) Definitions

Resistance (measured), R (unit: ohm).
Conductance (measured), C (reciprocal ohm, ohm^{-1}, or mho).
Specific resistance, ρ (ohm cm^{-1}).
Specific conductance, κ (ohm^{-1} cm^{-1}).
Equivalent conductance, Λ (ohm^{-1} cm^2 g.eq.$^{-1}$).

The reference in this definition to 'parallel electrodes 1 cm apart' causes much confusion. The relationship between specific and equivalent conductance should remove any difficulties. If the concentration of the electrolyte is c g.eq. per litre, then $1/c$ is the number of litres per g.eq., and $1,000/c$ is the number of cubic centimetres per g.eq. *If* each of these cubic centimetres is a one-centimetre cube, then

15

$1,000/c$ is the number of one-centimetre cubes containing 1 g.eq. of the electrolyte. The conductance of each of these is κ, and so

$$\Lambda = \frac{1,000\kappa}{c}$$

Note the distinction between 'number of one-centimetre cubes' and 'number of cubic centimetres'.

Limiting equivalent conductance, $\Lambda°$.
Ionic mobility (ionic conductance), λ, and its limiting value, $\lambda°$.
Absolute mobility of an ion, u (cm sec^{-1}).

Either of these last two terms may be referred to simply as the 'mobility' of an ion. The context will usually show which is intended, but the use of the easily distinguishable terms 'ionic conductance' and 'absolute mobility' is recommended, since λ is the contribution of the ion to the *equivalent* conductance of the solution.

Transport (or transference) number of an ion, t.
The relaxation or asymmetry effect in conductance.
The electrophoresis effect in conductance.
Ion-pair.

The distinction between incomplete ionization and incomplete dissociation is very fine. Ion-pairs can be formed in solutions of completely *ionized* electrolytes; they cause incomplete *dissociation*.

Reversibility of a galvanic cell.
Standard oxidation (or reduction) potential, $E°$.
Standard hydrogen electrode.
Liquid junction potential (or diffusion potential).
Activity, a, and mean ionic activity of an electrolyte, $a_{\pm}$.

The activity of a species is best defined in terms of the thermodynamic relationship between the chemical potential of the species under the specified conditions and its standard chemical potential.

Activity coefficient and mean ionic activity coefficient of an electrolyte.
Acidic and basic dissociation constants, K_A and K_B.
Definitions of acid and base according to Lavoisier, Davy, Ostwald and Arrhenius, Franklin, Brönsted and Lowry, and Lewis.
Hydrolysis constant of a salt, K_H.
The ionic product for water, K_W.
Degree of hydrolysis.
Ampholyte.
Indicator constant, K_I.
Buffer action.
Buffer capacity.

Decomposition potential.
Overpotential (or overvoltage).
Electrolytic polarization.

(ii) Laws and Equations

Faraday's two laws of electrolysis.
Kohlrausch's law (the law of the independent mobilities of ions).
Kohlrausch's equation.

> Avoid confusion between these two. Kohlrausch's law may be expressed as a simple equation, but the relationship known as 'Kohlrausch's equation' is quite different.

The Arrhenius relationship (conductance ratio).
The Debye–Hückel equation (for conductance).
The Onsager equation.
Walden's rule.
The Nernst–Thomson rule.
The Gibbs–Helmholtz equation (modified).
The Nernst equation.

> The original equation devised by Nernst for the potential of an electrode is obsolete, but the name is still applied to such equations in general. The potential of *any* electrode system can be expressed by means of one equation which involves the ratio of products of the activities of all components of the system in the oxidized state and in the reduced state. Attention must be paid to the *sign* of the term involving this ratio in accordance with the appropriate sign convention (*see* below).

The following equations for the e.m.f. of concentration cells, in terms of the mean ionic activities:
 Concentration cell with transport (or transference):
  (*a*) with anion electrodes;
  (*b*) with cation electrodes.
  Note particularly that the equation for *anion* electrodes contains the transport number of the *cation*, and vice versa.
 Concentration cell without transport.
  The use of mean ionic activities introduces the factor ν/ν_+ or ν/ν_- into the above equations. When 1:1 electrolytes are concerned this ratio is 2.
 Concentration cell with electrodes differing in concentration:
  (*a*) gas electrodes;
  (*b*) amalgam electrodes.
 Equations for:
  (*a*) transport number; and

(b) liquid junction potential
from concentration cells.

Equation for pH from the e.m.f. of a suitable cell.
The Henderson equation.
Equations for the hydrolysis constants of salts of a weak acid, a weak base, and both weak acid and weak base.
Equation for the degree of hydrolysis (from Ostwald's dilution law).
Equations for the pH of hydrolysed salt solutions.

(iii) Questions

(a) Transport Numbers

1. Review briefly the methods by which transport numbers may be determined and describe one method in detail. Show how such measurements may be used, in conjunction with conductance measurements, to determine ionic mobilities. (R)

2. Describe the Hittorf method for determining the transport numbers of the ions of an electrolyte. Show how the transport number of the hydroxide ion would be calculated from an experiment in which an aqueous solution of sodium hydroxide was electrolysed between platinum electrodes. (R)

3. Indicate the various methods by which the transport numbers of the ions in an electrolyte solution may be determined, and describe, with full experimental details, how one of these methods is carried out. (R)

4. Explain the terms (a) equivalent conductance, (b) ionic mobility and (c) transport number, and discuss the relationships between them. Outline very briefly the methods which can be used for the determination of transport numbers.

At 25° C the limiting equivalent conductances at infinite dilution of potassium chloride and sodium nitrate are 149·85 and 121·59 cm^2 ohm^{-1} equiv^{-1}, respectively. The transport numbers of the potassium and sodium ions in these solutions are 0·3906 and 0·4124, respectively. What are the limiting equivalent conductances of potassium nitrate and sodium chloride? (R)

5. Explain the following terms and show how they are inter-related: (a) equivalent conductance; (b) ionic mobility; and (c) transport number. Indicate methods which can be used to determine transport numbers and describe one method in detail. (S)

Notes. Questions on transport numbers follow a regular pattern, usually requiring a review of methods and/or details of one. Three general methods are available, namely, Hittorf's, moving boundary, and e.m.f. methods. The first of these is important historically and as a means of illustrating the meaning of transport number and its relationship to concentration changes, but it tends to over-emphasize the latter. Moving boundary methods, of which there are many modifications, are more popular and probably capable of greater accuracy. It can generally be said, however, that where there is a choice of methods one based on e.m.f. measurements, where applicable, will be the most sensitive because great accuracy can be attained in electrical measurements. This is so for the determination of transport numbers, but the method may be limited owing to a suitable cell not always being available or practicable. It is important to mention that the results of all three methods show very good agreement although based on entirely different principles.

Define and explain transport number even if not directly asked to do so, and quote some relationships with other electrochemical functions, particularly with equivalent conductance and ionic mobilities. This will help in answering other parts of the questions, including the importance of transport numbers.

The rider in question 2 requires careful thought. When considering the use of Hittorf's method for the transport number of a particular ion using a particular kind of electrode it is essential to take into account the reactions which will occur at the electrodes. In the present example the platinum electrodes are not attacked, i.e. they do not enter into the electrode reactions but merely serve to conduct electrons into and out of the solution. It is necessary to use dilute solutions, and the familiar gaseous products of the electrolysis of dilute solutions of acids and bases will be evolved. Consider carefully the equations for the electrode reactions, particularly the anode reaction. It will be necessary to assume that the base will suppress the dissociation of water to such an extent that the transport number of the hydrogen ion will be effectively zero, i.e. its migration may be neglected. If the changes near the two electrodes caused by migration and discharge of ions are tabulated it will be seen that the transport number of the hydroxyl ion may be expressed most simply in terms of the changes in its concentration near the cathode.

Question 4 includes a calculation which is a very simple application of Kohlrausch's law.

In addition to the relationships referred to above, transport numbers are important in relation to liquid junction potentials, and hence the use of salt bridges in galvanic cells, and also in polarography.

(b) Conductance

1. Define the terms conductance, specific conductance, and equivalent conductance, and show how they are related. Describe with full experimental details how the equivalent conductance of a solution can be determined. Indicate how the variation of equivalent conductance with concentration can be interpreted (a) for potassium chloride, and (b) for acetic acid. (R)

2. Explain what is meant by the *equivalent conductance* of an electrolyte. Discuss the interpretation of the variation of the conductance of an electrolyte with concentration (a) for the case of a strong electrolyte, and (b) for the case of a weak electrolyte. (R)

3. Given a pure anhydrous specimen of potassium bromide, describe the experiments you would carry out in order to determine accurately its equivalent conductance at zero concentration (infinite dilution) in water at 25° C. Mention the chief precautions you would take to avoid errors in the experimental work, and show how you would plot your results. Explain briefly the theoretical justification for the method of extrapolation you employ. (R)

4. Describe, with full experimental details, how you would proceed to ascertain the equivalent conductance of potassium bromide at infinite dilution in water. Outline the theoretical basis for the methods you would use in interpreting the experimental results and in extrapolating them. (R)

5. *Either* (a) Write an essay on conductometric analysis.

Or (b) Briefly explain the origins of the 'relaxation' and 'electrophoresis' effects according to the Debye–Hückel theory of conductance. Mention any experimental evidence for the existence of these effects.

6. State and explain Kohlrausch's Law of the Independent Mobilities of ions.

The mobilities of the silver ion and the chloride ion are 56·5 and 68·0 respectively at 20° C. A saturated solution of silver chloride in water at 20° C has a specific conductance of $1·37 \times 10^{-6}$ ohm^{-1}. Calculate the solubility of silver chloride at 20° C.

7. For the case of a strong 1:1 electrolyte show how the equivalent conductance is related to ionic mobilities (velocities). How can mobilities of individual ions be calculated from experimentally determined quantities? (S)

8. Discuss the application of electrical conductance measurements to *three* of the following:

(a) measurements of solubilities;
(b) study of colloidal electrolytes;
(c) determination of equivalence points in ionic titrations;
(d) measurement of dissociation constants of weak acids. (Ge)

9. Explain clearly the basic principles underlying the variation of the conductivity of a strong electrolyte with the concentration. Show how the thermodynamic dissociation constant of a weak electrolyte may be determined from conductivity measurements in dilute solutions. (Gi)

10. The limiting equivalent conductances of some ions in water at 25° C have the following values, in cm^{2}ohm^{-1}equiv^{-1}:

H$^+$, 349·81; Na$^+$, 50·10; Cl$^-$, 76·35; OH$^-$, 198·3.

100 cm^3 of a solution of 0·01N hydrochloric acid is titrated with 0·01N sodium hydroxide solution at this temperature. Calculate the specific conductances and the pH values after the addition of 50, 100 and 150 cm^3 of titrant. Define all the quantities you mention and explain any simplifying assumptions you make. (R)

See also questions S. 26; X. 20, 23.

Notes. The variation of conductance with concentration is a constantly recurring theme, of which the commonest variation is well set out in question 1. Where there is ambiguity, as in question 2, the advice given in Chapter 1 should be followed. The variation of conductance with concentration is usually interpreted in terms of the equivalent conductance, which is more generally useful in physical chemistry than is specific conductance. If any particular expression is required in terms of the latter this can readily be provided by means of the simple relationship which should have been mentioned in answering the first part of the question.

There are two factors involved in the interpretation of the concentration effects, namely, degree of dissociation, and ionic velocities. The first will be predominant in dilute solutions of weak electrolytes, whereas the speeds will be the deciding factors in solutions of strong electrolytes. Hence, the Arrhenius theory gives a reasonable interpretation of the variation of equivalent conductance of *weak* electrolytes with concentration. By combining the Arrhenius relationship with the Ostwald dilution law expression for the dissociation constant it is seen that the equivalent conductance is inversely proportional to concentration.

The relationship for strong electrolytes is more complex. An explanation of the asymmetry and electrophoresis effects of the Debye–Hückel theory is necessary. The equation, as modified by Onsager, shows that the equivalent conductance of a strong electrolyte is a linear function of the square root of the concentration. Here lies the explanation of the method of extrapolation of the graph when measuring the limiting conductance, i.e. the equivalent conductance at infinite dilution.

Temperature control and correction for the conductance of the solvent are necessary in accurate work. For the latter simple deduction is permissible when there is no reaction between the solute and any impurity present in the solvent, or the ions of the solvent itself.

The effect of ion-pair formation in solutions of strong electrolytes will be similar to those of incomplete dissociation, and allowance may be made for this by means of the Arrhenius expression and the Kohlrausch equation, or the Onsager equation, using the method of successive approximations. Remember that both the Kohlrausch and Onsager equations are limiting equations, applicable only in dilute solutions.

When describing experimental methods in electrochemistry it is essential to draw the electrical circuit, including that of any measuring instrument, simplified if necessary. This applies particularly to a conductance bridge; examiners like to be assured that students understand the working of such instruments and not merely which knobs to turn.

The calculation in question 6 is very simple if one assumes that the saturated solution is so dilute that its equivalent conductance does not differ appreciably from the limiting value.

Conductometric analysis includes direct conductometry besides conductometric titration.

In the second part of question 8 the use of conductance measurements to investigate the formation of micelles is required, including the determination of the critical concentration, micelle concentrations, and possibly also the effects of solubilization.

(c) Galvanic Cells

1. What is meant by a standard electrode potential?

Describe one electrode reversible to a cation, and one electrode reversible to an anion. Assuming that the two are combined to form a galvanic cell, write down the cell reaction, and explain what thermodynamic information the cell may be used to obtain. (R)

2. Explain the term *standard electrode potential* and, with reference to one example, indicate how it may be determined.

Derive an expression for the e.m.f. of a cell of the type Ag, AgCl|HCl (aq.)|(Pt) $\frac{1}{2}H_2$(1 atm.) and discuss the information that can be obtained from measurements on it. (R)

3. Describe with full experimental details how you would determine the e.m.f. of the cell

$$Zn|ZnSO_4\ (0\cdot5M)||CuSO_4\ (0\cdot5M)|Cu$$

in which the liquid junction potential is minimised by using a salt bridge between the solutions.

The e.m.f. of the above cell is 1·0947 volts at 12·0° C and 1·0818 volts at 42·0° C. Discuss the information which can be obtained from these measurements. (R)

4. Describe, with full experimental details, how you would determine the e.m.f. of a cell.

For the cell:

Pt, H_2 (1 atm.)|HCl (0·01M)||NaOH (0·01M)|H_2 (1 atm.), Pt

in which the liquid junction potential has been effectively eliminated by the use of a salt bridge, the e.m.f. is 0·585 V at 25° C. Show what information can be derived from this measurement. The mean ionic activity coefficients of hydrochloric acid and sodium hydroxide in 0·01M solution are each 0·905 at 25° C. (R)

5. Taking the cell:

$$H_2 \text{ (1 atm) } \overset{-}{Pt} \Big| \begin{matrix} 0\cdot1\text{M HCl} \\ 0\cdot1\text{M CuCl}_2 \end{matrix} \Big| \overset{+}{Cu}$$

as an example, state what is meant by the reversible E.M.F. of a galvanic cell, describe how it may be measured, and explain what information it may give about the cell reaction.

If, in the above example, the platinum and copper electrodes were electrically connected through a suitable resistance, describe the changes that would occur at the electrodes, and in the solution in the neighbourhood of each electrode. (R)

6. The galvanic cell:

$$H_2 \text{ (1 atm.), } \overset{-}{Pt}|\text{HCl (m)}|Hg_2Cl_2|\overset{+}{Hg}$$

has, at 25° C, the following values of E.M.F. for the given molalities, m, of hydrochloric acid:

m (mole kg^{-1})	0·07508	0·03769	0·01887	0·00504
E (volt)	0·4119	0·4452	0·4787	0·5437

By a simple graphical method determine the standard E.M.F. of this cell and calculate the activity coefficient of hydrochloric acid in the most concentrated of the four solutions mentioned. (R)

7. Explain what is meant by oxidation–reduction potential and derive an expression for the oxidation–reduction potential of the ferric–ferrous system.

Discuss briefly the use of oxidation–reduction potentials in *either* (*a*) the determination of pH, *or* (*b*) quantitative analysis. (R)

8. Explain, with examples, what is meant by a reversible cell. Show how measurements of the e.m.f. of reversible cells can give information of physico-chemical interest. (Ge)

9. Explain the concept of activity. Show how it is applied to the determination of the standard electrode potential of an electrode system of your own choice. (Ge)

10. By application of thermodynamic reasoning to the reaction

$$\tfrac{1}{2}H_2(gas) + Fe^{3+}(aq) = H^+(aq) + Fe^{2+}(aq)$$

derive an expression for the e.m.f. of the cell

$$H_2, Pt|H^+(aq)||Fe^{3+}, Fe^{2+}|Pt$$

as a function of hydrogen gas pressure, hydrogen ion activity, ferric ion activity and ferrous ion activity. Any liquid junction potential is to be ignored. Indicate briefly how you would set up such a cell in the laboratory. (Gi)

11. What do you understand by 'oxidation–reduction potential'? By means of thermodynamic arguments derive an expression for the oxidation–reduction potential of the ferric–ferrous system. Outline *either* the application of an oxidation–reduction electrode to the determination of pH, *or* illustrate the usefulness of oxidation–reduction potentials in relation to analytical reactions. (S)

12. Explain how measurements of the E.M.F. of appropriate electrochemical cells may be used to determine the standard changes in Gibbs free energy, entropy, and enthalpy at 25° C for the reaction

$$H^+(aq) + OH^-(aq) = H_2O\ (l)$$

(S)

13. Define the terms 'standard oxidation potential' and 'standard reduction potential'. Explain, with essential experimental details, how you would measure, by means of a suitable electrochemical cell, the standard oxidation potential of the ferrocyanide–ferricyanide system in aqueous solution at 25° C. Summarise briefly the importance of a knowledge of such potentials. (S)

14. Devise an electrochemical cell which could be used to determine the standard Gibbs free energy change at 25° C of the reaction

$$H_2(g,\ 1\ atm.) + 2Hg^{2+}(aq) \longrightarrow 2H^+(aq) + Hg_2^{2+}(aq).$$

Explain the experimental procedure and calculations involved. (S)

15. Show, with examples, how e.m.f. measurements on suitable cells can yield thermodynamically important information about solutions of strong electrolytes. (Gi)

16. Discuss the application of e.m.f. measurements to *one* of the following problems, giving essential experimental details:

(*a*) The determination of $\Delta G°$ and $\Delta H°$ for the reaction:

$$H_2(g) + 2\ AgCl(s) \rightleftharpoons 2HCl(aq) + 2Ag(s).$$

(*b*) The determination of activity coefficients. (Gi)

17. *Either* Discuss the significance in chemistry of standard electrode potentials, and outline how they may be determined.

Or Define the term 'concentration cell without transport'. Give an example of such a cell, derive an expression for its e.m.f. and explain how it may be applied to the evaluation of activity coefficients. (Gi)

18. In what galvanic cell is the reaction:

$$\tfrac{1}{2}H_2 + AgCl = Ag + HCl(aq)$$

the source of the e.m.f.? Outline the construction of such a cell and the method used to measure its e.m.f. in order to obtain results of thermodynamic significance. Derive an expression for the e.m.f. of this cell by means of thermodynamic reasoning. The e.m.f. of the cell with H_2 pressure = 1·000 atm. and HCl molality of 0·1000 mole kg^{-1} is 0·3524 abs. volt at 25° C. The mean ion activity coefficient of HCl in this solution is 0·796. Calculate the standard e.m.f. of the cell, $E_m°$.

19. For the cell:

$$H_2 \text{ (1 atm.) } \overset{-}{Pt}|HCl \text{ (aq)}|Hg_2Cl_2|\overset{+}{Hg}$$

write down the cell reaction, and from basic thermodynamic principles deduce a relationship between the e.m.f. of the cell and the mean ionic activity of the hydrochloric acid.

At 25° C. the e.m.f. of this cell when the acid is 0·01 molal is 0·5098 v., and the standard e.m.f. is 0·2685 v. Calculate the mean ionic activity coefficient of hydrochloric acid at this concentration, and compare it with the value obtained from the limiting Debye–Hückel law (at 25° C. the constant, A, is 0·509). Briefly indicate why the use of the limiting law should be restricted to very dilute solutions. (Ge)

20. Explain how *two* of the following can be determined from measurements of the e.m.f. of suitably chosen cells (all in aqueous solution):

(a) the solubility product of silver bromide;
(b) the dissociation constant of benzoic acid;
(c) the mean ionic activity coefficient of sodium hydroxide;
(d) the equilibrium constant for the reaction:

$$Ce^{4+} + Fe^{2+} \rightleftharpoons Ce^{3+} + Fe^{3+}$$

(R)

21. For the following cell reactions (a) give the component electrode reactions, (b) draw the cell diagram, (c) calculate the Standard E.M.F. of the cell, (d) show which is the negative electrode, and (e) calculate the equilibrium constant for the *spontaneous* cell reaction:

(i) $Fe + 2Fe^{3+} = 3Fe^{2+}$
(ii) $VO_2^+ + V^{3+} = 2VO^{2+}$
(iii) $H_2 + Cl_2 = 2H^+ + 2Cl^-$
(iv) $Ag_2SO_4 + Pb = PbSO_4 + 2Ag$

A list of Standard Electrode (Reduction) Potentials is provided. (D)

Notes. A standard electrode potential is exactly what it says, viz. the potential of an electrode system when all the component substances are in their standard states. Define the latter for the three states of matter, and explain how, by convention, the actual values of electrode potentials used are relative to that of the standard hydrogen electrode, since absolute values are not available. The measurable quantity is an e.m.f., which is the algebraic sum of all the potential differences arising in a cell. This relationship between e.m.f. and electrode potentials must be clearly brought out and leads to the explanation of a sign convention for both. Most examining bodies now recommend the use of the IUPAC sign convention (*see* Chapter 4).

Refer to Section 1.5 for advice on the correct choice of examples. An additional point here is that it is safer to select first a cell with which you are familiar and describe the anion and cation electrodes used. An arbitrary choice of examples of electrodes could lead to difficulties when discussing the information available from the cell. Convenient examples would be a silver (cation) electrode and a silver–silver chloride (anion) electrode. When suitably combined the cell reaction would be the ionic dissociation of silver chloride. Such a cell could be used to determine the usual thermodynamic functions for the reaction, including the equilibrium constant, which will be the solubility product of silver chloride. Alternatively, the electrodes of the cell mentioned in question 2 could be used as examples.

Occasionally one section of a question partly answers another section. This is illustrated in question 2. The cell referred to is quite suitable as an example to demonstrate the method of determining a standard electrode potential. Also, the latter is an example of the information which can be obtained from measurements on the cell, as required for the last part of the question. There is, of course, other information available also.

Experimental details must include an electrical circuit and a drawing of the arrangement of the cell, particularly when a salt bridge is to be used. Selection of a suitable salt for the bridge is important. In order to minimize the liquid junction potential the salt must have a common ion with each solution, and the transport numbers of its ions must be as nearly equal as possible. This points to ammonium sulphate in question 3 and sodium chloride in question 4, but it should be emphasized that the presence of hydrogen or hydroxyl ions tends to produce large liquid junction potentials, and it is doubtful whether the arrangement suggested in question 4 would be entirely satisfactory.

In order to discuss the information available from any given measurements it is essential to carry out the appropriate calculations.

The cell referred to in question 4 is a concentration cell consisting of two hydrogen electrodes (with the gas at the same pressure) in solutions of different hydrogen ion activities. In the alkaline solution the activities of the hydrogen and hydroxyl ions are related by the ionic product for water. The cell can therefore be used to determine this.

For a 1:1 electrolyte the individual ion activity coefficients may be taken as equal to the mean ionic activity coefficient of the electrolyte, i.e.

$$f_+ = f_- = f_\pm$$

Reversibility in a galvanic cell (questions 5 and 8) implies both chemical and thermodynamic reversibility and is explained in terms of the effects of applying an opposing e.m.f. The last part of question 5 is rather unusual, and a very good question for revision purposes. Changes occurring at the electrodes as the cell is discharging, and the e.m.f. falling towards zero, are shown by the equations for the electrode reactions. Those occurring in the solution near the electrodes may be deduced by considerations of discharge, deposition, or dissolution of ions, in addition to the transport of ions through the solution under the potential gradient.

The graph to be plotted in answer to question 6 will be $(E_{cell} + 2 \times 0.05916 \log m)$ against $m^{\frac{1}{2}}$. Extrapolation of this to zero concentration gives the standard e.m.f. It will be necessary to derive the expression on which the graph is based. Then, by inserting in this equation the values relating to the most concentrated solution, calculate the mean activity coefficient of hydrochloric acid. An explanation of the procedure is also required for question 9.

It must be emphasized that oxidation and reduction may take place in any reversible electrode system or half-cell, but the expression 'oxidation–reduction potential' is reserved for systems in which the oxidized and reduced states are present *together* in the solution. This is in contrast to many other systems where one or more species in the oxidized or reduced states (or both) are solids. The platinum of the common form of an oxidation–reduction electrode is *not* part of the electrode system and does not enter into the electrode reaction. It merely makes electrical contact with the system.

The discussion in question 7 will relate to oxidation–reduction indicators, systems in which hydrogen ions are involved, such as the 'quinhydrone electrode', and the use of other oxidation–reduction systems in potentiometric titrations.

Activity (question 9) must be defined thermodynamically by reference to the chemical potential. It is often considered as an idealized concentration, but, more correctly, one should regard concentration as an approximation for activity, which is the fundamental function. The explanation of activity will follow logically from this approach. Note that e.m.f. methods always involve activities, but that for analytical purposes results have to be given in terms of concentrations. The conversion is not always without its difficulties.

University examiners place a strong emphasis on the thermodynamic approach to electrochemistry, as illustrated again in question 10. The derivation will be through the thermodynamic equilibrium constant for the given reaction.

In question 14 the cell required will clearly consist of a hydrogen electrode and a mercurous/mercuric oxidation–reduction electrode. The procedure for finding the standard e.m.f. has already been mentioned, and the standard Gibbs free energy change follows directly. In selecting suitable components for the cell one must consider the solubility and degree of dissociation of the mercury salts, and choose a suitable salt bridge on the basis discussed above (questions 3 and 4).

(d) pH

1. Discuss, with particular reference to the principles involved, methods available for the determination of the pH of a solution. (R)
2. Give a critical account of methods which can be used to determine the pH of a solution, indicating conditions under which the various procedures would be favoured. (R)
3. A soluble monobasic acid of dissociation constant $K = 2\cdot 0 \times 10^{-4}$ moles per litre is half neutralised with sodium hydroxide. Calculate the approximate pH of the solution produced, and mention any reasons for regarding your answer as being only an approximate one.
   Describe one method for determining the pH of a solution. (R)
4. Discuss the measurement of hydrogen ion concentration by electrometric methods and the control of hydrogen ion concentration by the use of buffers. (Ge)

Notes. Of the very few general methods for the determination of the pH of a solution only the colorimetric and electrometric methods are commonly used. The catalysis method is of purely historic interest but examiners appear to appreciate a brief mention of it. On the other hand, there is considerable variety in the detailed application of the methods, as, for example, in the kinds of electrodes used with pH meters. The basic circuits of the two main types of pH meter must be described along with the relationship between pH and e.m.f.

An outline of indicator theory is required in order to explain the principles of the colorimetric method for pH, explaining, in terms of the indicator constant, why an indicator changes colour over a definite range of pH and why the ranges of various indicators differ.

Question 3 is a straightforward application of the Henderson equation. As the actual concentrations of the acid and salt are not given, only the simplified form of the equation may be used. Much time may be saved if the significance of the statement 'half neutralized' is appreciated. The reasons for the approximate nature of the result may be shown only by use of the full equation, which should be deduced.

Electrometric methods, i.e. pH meters, measure the hydrogen ion *activity*, so that the discussion in the last question must refer to the relationship between concentration and activity. Neutralization curves and the Henderson equation will explain buffer action and buffer capacity.

(e) *Acids and Bases*

1. Define, and comment briefly on, the terms *acid* and *base*.

The acid dissociation constant of the pyridinium ion, $C_5H_5NH^+$, is $4\cdot 35 \times 10^{-4}$ in water at 25° C. Calculate the degree of hydrolysis of a 0·1N solution of pyridine hydrochloride, and the basic dissociation constant of pyridine in water. The ionic product of water is $1\cdot 0 \times 10^{-14}$. (R)

2. Write a very short essay on acids and bases. The ionic product of water, $K_W = [H^+][OH^-]$, where the terms in brackets represent activities, has the values of $0\cdot 681 \times 10^{-14}$ at 20° C and $1\cdot 471 \times 10^{-14}$ at 30° C. Assess the heat of neutralization of one equivalent of a strong acid by a strong base at 25° C. (R)

3. Write a short essay on *either*

  (*a*) Adsorption; *or*
  (*b*) The nature of acids and bases. (R)

4. Give an account of current theories of acids and bases. (R)

5. Give an account of *one* method for the accurate determination of the dissociation constant of a weak acid in aqueous solution. Comment briefly on the relationship of the second dissociation constant to the first dissociation constant in the case of a dibasic acid. (Ge)

6. Write an account of:

  Either (*a*) the interpretation of acid–base phenomena in aqueous and non-aqueous solvents in terms of the Brønsted–Lowry concept,
  Or (*b*) the principles of methods available for studying the formation, formulae and stability constants of complex ions in solution. (Ge)

7. Write a concise account of acid–base catalysis from the standpoint of the modern interpretation of the terms 'acid' and 'base'. (D)

See also questions X. 17, 30.

Notes. The most generally useful definitions of acid and base are those of the Brønsted–Lowry theory. These will suffice where a straightforward definition is asked for, but when reference is made to current theories the Lewis definition must also be given. In a general essay on acids and bases it might be useful to adopt the historical approach, showing the development of the definitions, but giving more time to detailed discussion of the applicability of the two current theories. At the same time the solvent-system theory of Franklin should not be overlooked.

The calculation in question 1 should present no difficulty provided that the dissociation constants are derived on the basis of the Brönsted–Lowry theory. Calculations of the kind given in question 2, involving the use of the integrated form of the van't Hoff isochore, often end in confusion if incorrect signs are used or one fails to take the logarithm of $\log(K_1/K_2)$.

Sulphuric acid is an example to illustrate the universal relationship between the first and second dissociation constants of a dibasic (or tribasic) acid. The second constant is always less than the first.

(f) *Miscellaneous*

1. Write on *two* of the following:

(a) The selection of indicators for acid–base titrations;
(b) Buffer solutions;
(c) The hydrolysis of salts;
(d) Solubility product. (R)

2. Give an account of the theory underlying *either* (a) potentiometric titrations, *or* (b) conductometric titrations. Describe the procedure for the technique chosen and how the titration curves should be interpreted. (R)

3. Write notes on *two* of the following:

(a) buffer solutions;
(b) redox indicators and their uses in titrations;
(c) the determination of the structure of the sodium chloride crystal;
(d) the determination of the charge on the electron. (R)

4. Outline the principles involved in one method for each of *two* of the following determinations:

(a) the transport number of an ion in a solution of a simple binary electrolyte;
(b) the solubility product of mercurous chloride;
(c) the critical volume of carbon dioxide;
(d) the pH of a buffered solution. (Ge)

5. Write briefly on *three* of the following:

(a) the definition of pH and the standardisation of the pH scale,
(b) the choice of indicators for acid–base titrations,
(c) buffer action,
(d) Hammett acidity functions. (Gi)

6. Write a short essay on the non-ideal behaviour of *fully ionised* solutes in aqueous solution. Avoid any detailed mathematical derivations but pay attention to the definition of standard (ideal) states and the methods by which the extent of non-ideality is represented and measured. (S)

7. Write brief notes on any *two* of the following topics:

(a) The determination of ionic mobilities;
(b) The conductimetric determination of the dissociation constants of weak acids;
(c) Potentiometric titrations;
(d) Polarography. (Gi)

8. *Either* explain briefly the principles of *two* of the following determinations, *or* choose *one* of them for more detailed discussion of theory, experimental procedure and accuracy:

(a) determination of the transport number of the hydrogen ion in dilute aqueous HCl solution;
(b) determination of the pH of a solution using *either* the quinhydrone electrode *or* the antimony electrode;
(c) determination of an ionisation constant by use of a spectrophotometric method. (Gi)

9. Discuss potentiometric methods by which *three* of the following may be determined:

(a) the pH of an aqueous solution;
(b) the dissociation constant of a weak acid;
(c) the dissociation constant of water (K_W);
(d) transport numbers. (R)

10. Outline the method of Potentiometric Titration.

Consider the practicability of studying the kinetics of the following reactions by potentiometric methods (not titrations):

(a) $2KI + K_2S_2O_8 = I_2 + 2K_2SO_4$
(b) $CH_3COOC_2H_5 + NaOH = C_2H_5OH + CH_3COONa$ (D)

See also questions S. 24; T. 44; X. 18, 19, 25, 26, 27.

Notes. When writing notes on chosen topics it is very important to include sufficient detail and examples to show that one's knowledge is not merely superficial. Question 1 is rather vaguely phrased, but it is clear that in the fifteen or so minutes available for each section a full essay cannot be satisfactorily written. Refer to comments on 'notes' questions in Section 1.3.

Potentiometric titration curves are explained in terms of the equation for the potential of the indicator electrode and the Henderson equation. The latter should be derived and the applicability of the full and simplified forms discussed. The interpretation of the titration curves should include the normal curve of e.m.f. against volume of reagent added and the first and second derivative curves, with indications of the equivalence point in each case. The method of calculation of the end-point by interpolation should be explained, i.e. to find the volume of reagent at which the second derivative of e.m.f. with respect to volume added is zero.

In conductometric titrations a consideration of the titration reaction will show the ionic changes occurring. The type of curve produced will depend on changes in both the concentrations of the ions and the nature of the ions remaining in solution, the latter owing to the different speeds of various ions. Addition or removal of hydrogen or hydroxyl ions will have the greatest effects.

Variations in speeds of ions are explained in terms of the Debye–Hückel theory, and relative speeds in terms of ion sizes and extents of hydration.

K. CHEMICAL EQUILIBRIUM AND KINETICS

A knowledge of the simple kinetic theory of gases is essential to an understanding of reaction kinetics, and particularly of the collision theory. For this reason kinetic theory is included in this section.

Radiation chemistry is also included here since it is merely the study of reactions initiated by the absorption of ionizing radiation, though few questions arise at this level. A distinction might be made between photochemistry, in which the radiation absorbed is in the range of what is usually called optical spectroscopy, and radiation chemistry, in which the radiation is generally of nuclear origin or X-rays.

(i) Definitions

Active mass.
Mean square velocity.
Mean velocity.
Most probable velocity.

These three velocities must not be confused. A simple example will illustrate the differences. Suppose we have three particles with velocities 2, 4, and 6 cm sec^{-1}. The mean velocity is clearly 4 cm sec^{-1}, and the square of the mean velocity is 16 cm^{2}sec^{-2}. But the mean square velocity, $\overline{c^2}$, is $(2^2 + 4^2 + 6^2)/3$, which is 18·67 cm^{2}sec^{-2}. The most probable velocity, c, is $(2RT/M)^{\frac{1}{2}}$ and can be shown to be equal to $(2/3)^{\frac{1}{2}} \times (\overline{c^2})^{\frac{1}{2}}$, i.e. 3·53 cm sec^{-1}, and c^2 is 12·45 cm^{2}sec^{-2}.

K. CHEMICAL EQUILIBRIUM AND KINETICS

Mean free path.
Collision diameter.
Collision number.
Order of a reaction.
Molecularity of a reaction.

    It is extremely important to distinguish these. Order of reaction refers to concentration terms, whereas the molecularity refers to numbers of particles and thereby implies a knowledge of the mechanism of the reaction.

Isolated reaction.
Rate constant (velocity constant or specific reaction rate, etc.).
Time of half-change (or half-life).
Pseudo-unimolecular reaction.
Zero order reaction.
Autocatalysis.
Fractional order reaction.
Induction period.
Chain initiation, propagation, and breaking.
Atom chain, radical chain, and energy chain.
Explosion limits.
Stationary chain.
Activation energy.
Frequency factor.
Probability or steric factor.
Activated complex or transition state.

    An activated complex is a short-lived, loosely bound intermediate in a reaction; it must not be confused with an activated molecule, which differs from a normal molecule *only* in having excess energy.

Partition function.
Molar extinction coefficient.
Absorption coefficient.
Resonance fluorescence.
Quantum efficiency of a photochemical reaction.
Photosensitization.
Chemiluminescence.

(ii) Laws and Equations

Law of mass action.
Le Chatelier's principle.
Equation for PV from the simple kinetic theory.
Equation for the mean square velocity.
Kinetic equation of the first order.

34 ANSWERING QUESTIONS IN PHYSICAL CHEMISTRY

Kinetic equation of the second order.
Simplified kinetic equation of the second order, for equal initial concentrations of reactants.
Kinetic equation of the third order.
General expression for the time of half-change.
Kinetic equation for a reversible reaction, of the same order in each direction.
Kinetic equation for consecutive reactions of the first order.
Kinetic equation for simultaneous reactions.
Langmuir's equations for a heterogeneous gas reaction (one or two gases).
The Arrhenius equation.
Hinshelwood's equation for the rate of a chain reaction.
Formulae for calculating the various partition functions.
The Grotthus–Draper law.
Lambert's law.
Beer's law.
The Franck–Condon principle.

(iii) Questions

1. Give an account of the simple kinetic theory of gases and show how it leads to the ideal gas laws. Indicate reasons for departures from these laws and discuss briefly how they are allowed for in the van der Waals equation. (R)

2. Discuss the factors which influence the velocities of chemical reactions. (S)

3. Explain what is meant by the order of a reaction, and describe methods by which it may be deduced from suitable experimental data. Discuss the variation of the rate constant of a reaction with temperature. (R)

4. Distinguish clearly, with examples, between the order and the molecularity of a reaction.

Explain *three* of the following terms which are used in discussing the rates and mechanisms of chemical reactions:

    (a) activation energy;
    (b) transition state;
    (c) chain-branching;
    (d) steady-state hypothesis. (Ge)

5. Give concise definitions of:

    (a) the kinetic order; and
    (b) the molecularity of a chemical reaction.

Derive criteria for first and second order reactions. (Gi)

K. CHEMICAL EQUILIBRIUM AND KINETICS

6. What do you understand by 'first order reaction'? Define the term 'velocity constant' and indicate the units in which the velocity constant of a first order reaction must be expressed. Give an example of a first order reaction, indicating the experimental method used to determine its rate. Show how the velocity constant for this reaction may be derived from the results (a) by a graphical method and (b) by the 'half-change' method. (R)

7. From the following data, infer the kinetic order and calculate the velocity constant for the thermal decomposition of the gas A:

time (min):	30	53	100
%A decomposed:	32	50	73

What further experiments might distinguish between the hypotheses (a) that the process is a chain reaction; (b) that it is unimolecular? (Gi)

8. State clearly what is meant by the order of a chemical reaction and show how it can be determined from suitable experimental data.

Explain why:

(a) the hydrolysis of an ester in the presence of dilute acid follows first-order kinetics whilst that in the presence of dilute alkali follows second-order kinetics;

(b) the combination of hydrogen and bromine does not follow simple second-order kinetics. (R)

9. The hydrolysis of ethyl acetate in aqueous solution is catalysed by hydrogen ions. Explain, with essential experimental details, how you would determine (a) the order of reaction, (b) the rate constant, and (c) the activation energy. (S)

10. Describe one method by which the order of a gas reaction might be determined. If the reaction is a reversible one show how the rate constants of the forward reaction and the opposing reaction can be evaluated.

What is the energy of activation of a reaction, and how can it be determined? (R)

11. The decomposition of gaseous 2-chloropropane into propylene and hydrogen chloride is of the first order at 350–400° C. Explain, with essential experimental detail, how you would (a) measure the frequency factor and activation energy, and (b) verify that the reaction is homogeneous and non-chain. (S)

12. Explain concisely the meaning of *three* of the following:

(a) transition state;
(b) branching and non-branching chain reactions;
(c) primary kinetic salt effect;
(d) quantum yield of a photochemical reaction.

In each case give *one* specific example to illustrate your explanation. (S)

13. A certain ester is fairly soluble in water, and its hydrolysis is found to be catalysed by hydrogen ions. Describe with experimental details how you would determine:

    (a) the rate constant under specified conditions;
    (b) the order of the reaction with respect to each reactant;
    (c) the energy of activation of the reaction. (R)

14. The iodination of a ketone $R_2CH \cdot CO \cdot CR_3$, where R is an alkyl group, to give $R_2CI \cdot CO \cdot CR_3$ in aqueous mineral acid is of the first order with respect to the ketone and of zero order with respect to iodine; the rate is also proportional to the acid concentration. When the initial concentrations are ketone 0·001M, iodine 0·002M, and acid 0·1M, the first order rate constant is 0·01 sec.$^{-1}$. What fraction of the iodine would remain after 2000 sec. if the acid concentration was 0·05M and the other concentrations were as stated above?

Discuss a mechanism for the reaction. (S)

15. *Either* give a general account of the determination and significance of the order and temperature coefficient of a chemical reaction;

    or discuss the theoretical interpretation of the pressure dependence of the kinetic order of a gaseous unimolecular reaction. (Gi)

16. Discuss the terms (a) reaction velocity constant and (b) activation energy.

Derive the integrated rate equation for a first order reaction.

Describe *briefly* how you would determine whether a reaction was kinetically of the first order. (R)

17. The rate constant of a chemical reaction can often be represented by an expression of the type $A\exp(-E/RT)$. Discuss the meaning of the terms A and E for (a) a bimolecular reaction in the gas phase, and (b) a unimolecular reaction in the gas phase. (Ge)

18. The velocity constant (k) of a bimolecular reaction may be approximately represented by the expression $k = A\exp(-E/RT)$. Explain the significance of A and E, (a) in terms of collision theory, and (b) in terms of transition state theory. Explain briefly why the latter has been more successful in the interpretation of reactions in solution. (Gi)

19. How is 'the activation energy' of a chemical process determined? Explain its significance in theories of chemical reactivity.

In a reversible reaction, what is the relationship between the activation energies of the forward and back reactions? (R)

20. The conversion of N-chloroacetanilide into *p*-chloroacetanilide can be followed by utilising the fact that the former liberates iodine from potassium iodide whereas the latter does not. Describe how you would

investigate the effects on the rate of the reaction of (a) the concentration of N-chloroacetanilide, and (b) the concentration of hydrogen ion. (R)

21. Describe with full experimental details how you would study the kinetics of any particular reaction of your choice, and how you would proceed to interpret the results of your measurements. (R)

22. The reaction

$$CH_2Br \cdot CO_2^- + S_2O_3^{2-} = CH_2(S_2O_3)CO_2^{2-} + Br^-$$

can be followed in aqueous solution by iodine titrations. Describe the experiments you would carry out to study the kinetics of the reaction, and explain fully how you would use your results to calculate the second-order rate constant. (R)

23. Give an account of *one* of the following topics:

(a) the kinetics of reactions involving ionic reactants in aqueous solution,

(b) the characteristics of heterogeneous reactions,

(c) the use of kinetics in the study of reaction mechanisms. (Ge)

24. *Either* give an account of unimolecular and bimolecular reactions in the gas phase, *or* give an account of the catalysis of gaseous reactions by solid surfaces. (Ge)

25. Give an account of the concept of the activated complex in reaction kinetics. (Gi)

26. Explain, with examples, the meaning and significance of the following terms in the theory of chain reactions: (a) initiation, (b) propagation, (c) termination, (d) branching. (Ge)

27. Give an account of photochemical reactions with particular reference to general principles which have been useful in the interpretation of the observed phenomena. (Ge)

28. Describe a method by which the gaseous equilibrium $N_2O_4 \rightleftharpoons 2NO_2$ could be investigated. How could values of K_p, the equilibrium constant in terms of partial pressures, be evaluated and the heat of dissociation at constant pressure be derived? (R)

29. Discuss the Arrhenius Theory of reaction kinetics and explain the significance of all the terms in the Arrhenius Equation.

The first order decomposition of formic acid on a gold surface has specific rate constants of $5 \cdot 5 \times 10^{-4}$ sec^{-1} and $9 \cdot 2 \times 10^{-3}$ sec^{-1} at 140° C and 185° C respectively. Calculate the values of the two constants in the Arrhenius Equation. Comment on the values obtained. (D)

30. (a) The reaction:

$$2NO + 2H_2 = N_2 + 2H_2O$$

was studied with equimolecular quantities of nitric oxide and hydrogen at various initial pressures. Calculate the order of the reaction graphically from the data given:

Initial pressure (mm)	354	340·5	375	288	251	202
Time for completion of half the reaction (min)	81	102	95	140	180	224

(b) The inversion of sucrose:

$$C_{12}H_{22}O_{11} + H_2O = C_6H_{12}O_6 + C_6H_{12}O_6$$

proceeded as follows at 25° C.:

Time (mins.)	0	30	60	90
Sucrose changed (moles/litre)	0	0·1001	0·1946	0·2770

The initial concentration was 1·0023 moles/litre. Calculate the first order reaction rate (mean value for observations given). How long would it take for 95% of the sugar to be inverted? Why is the reaction first order, when the equation shows the reaction to take place between two molecules? (D)

31. Explain, using a simple mathematical treatment, the conditions for an explosion in a chain reaction and shew why such reactions may exhibit an upper and a lower explosion limit.

Give a brief description of the possible mechanism and main features of *either* (a) the reaction $2H_2 + O_2 = 2H_2O$ *or* (b) organic polymerisation reactions. (D)

32. Illustrate and explain *two* of the following:

(a) A unimolecular gas reaction, taking place on the walls of the containing vessel, may, under appropriate conditions, have any order from zero to unity.

(b) Although some homogeneous gas reactions are believed to be unimolecular, the mechanism still supposes binary collisions as a necessary step in molecules acquiring sufficient activation energy to react.

(c) The function of surfaces in heterogeneous catalysis. (D)

33. From the Kinetic Theory derive an expression for *either*

(a) the viscosity of a gas; *or*
(b) the rate of diffusion of a gas.

Also, define and give expressions for the following:

(1) Mean Free Path
(2) Collision Diameter
(3) Collision Number
(4) Rate of effusion of a gas. (C)

K. CHEMICAL EQUILIBRIUM AND KINETICS

34. Discuss *three* of the following:

(a) The apparent order of the homogeneous para-hydrogen conversion at high temperatures is 1·5.

(b) Few homogeneous third-order gas reactions are known, and all of these involve nitric oxide.

(c) Atom and radical combination reactions frequently require the participation of a 'third body'.

(d) It is improbable that the reaction between stannous and ferric salts in solution really proceeds according to the equation

$$2Fe^{3+} + Sn^{2+} = 2Fe^{2+} + Sn^{4+} \qquad (C)$$

35. Briefly explain *two* of the following:

(a) The Lindemann theory of unimolecular reactions.
(b) Langmuir's theory of gas reactions at solid surfaces.
(c) The Activated Complex theory.
(d) Hinshelwood's equation for the rate of a chain reaction. (C)

36. Define the terms 'Order' and 'Molecularity' as applied to chemical reactions.
Answer *both* of the following:

(a) Half-change times for the thermal decomposition of nitrous oxide were found to be 255 sec and 212 sec for initial pressures of 290 and 360 mm respectively. Assess the order of the reaction.

(b) In the saponification of 0·01 molar methyl acetate by 0·01 molar sodium hydroxide at 25° C the following results were obtained:

t	180	300	420	600	900
x	2·60	3·66	4·50	5·36	6·37 × 10⁻³ mole/litre.

Show graphically and by calculation that the reaction is of the second order. Determine the rate constant and state the units. (C)

37. Outline the Lindemann Theory of unimolecular gas reactions. How does this theory explain the kinetics of bimolecular gas reactions at low pressures? (C)

38. Why is the 'order' of a chemical reaction not always the same as the 'molecularity'? In what kind of reaction are the two necessarily equal?

In the decomposition of N_2O_5 in carbon tetrachloride at 30° C, the volume of oxygen produced at any time is V, and the volume of oxygen on completion of the decomposition is V_∞. From the following results find, graphically or otherwise, the order of the reaction and calculate the velocity constant, stating the units:

Time (secs.)	0	2400	4800	9600	14400	∞
$(V_\infty - V)$ (cc.)	84·85	69·20	57·20	39·00	26·55	0·00

(C)

39. *Either:* What is a chain reaction? State three criteria by which such a reaction may be recognised, and discuss the occurrence of upper and lower 'explosion limits' in gaseous reactions.

Or: Describe an experiment for the determination of the energy of activation of a reaction, indicating clearly the method of calculation of the result. (C)

See also questions S. 25, T. 12, and X. 32.

Notes. One or two common errors might be noted here. It is vital to distinguish carefully between K (the equilibrium constant) and k (the rate constant). The variation of the former with temperature is expressed by the van't Hoff isochore, whereas the variation of rate constant with temperature is the basis of the Arrhenius theory of reactions and defines the activation energy. Furthermore, although the *rates* of the forward and reverse reactions become equal at equilibrium the *rate constants* clearly do not, since they are *constants* (at a fixed temperature) and $K = k_1/k_2$.

A certain amount of confusion often appears in connection with the time of half-change of a reaction ($t_{\frac{1}{2}}$). This is exactly what its name implies, namely, the time for the reaction to be half completed. It is quite different from 'half the time for the complete reaction'. Consideration of a graph of concentration against time for a first-order reaction will quickly show the difference between these two terms, and, indeed, the simple ratio between them.

Complaints of vagueness in answers are more frequent in kinetics than elsewhere. Most students have little difficulty in understanding the theoretical aspects of reaction kinetics, but have trouble over practical methods and particularly in the choice of examples. The advice of Chapter 1 should be taken here. During revision write down the various kinetic equations and under each give brief details of a suitable example, i.e. a reaction which follows the equation concerned, does not involve any side issues, and preferably is an elementary reaction. Consider the possibility of using free-radical reactions in some cases, and especially in explaining the activation energy. Note also that the use of 'molecularity' implies that the *mechanism* of the reaction is known.

The units in which rate constants are expressed must be clearly stated. There should be no difficulty in arriving at the correct units by combining the units of concentration and time with the kinetic equation used. Thus, for a second-order reaction the value of k is generally expressed (in the simplified form of the equation) as

$$k = \frac{1}{t} \cdot \frac{x}{a(a-x)}.$$

With one concentration term in the numerator and the product of two concentration terms in the denominator we have (concentration units)$^{-1}$. Bringing in the time factor we arrive at (concentration units)$^{-1}$(time units)$^{-1}$, for example, (moles per litre)$^{-1}$sec^{-1}, or l.mole^{-1}sec^{-1}. The time units (usually seconds or minutes) are just as important as the concentration units.

The individual questions require little further comment. In the last part of question 4 do not confuse 'transition state' and 'the steady-state hypothesis'. It is emphasized again that the former is also known as the 'activated complex' and is a short-lived, loosely bound complex, intermediate between reactants and products. The steady-state hypothesis (or approximation) refers to relatively short-lived intermediate species formed in chain propagation reactions. The mathematical treatment is simplified by assuming that the concentration of such intermediates remains constant, i.e. they are in a 'steady state'.

Question 7 involves the criteria of chain reactions and of unimolecular reactions. The last part of the question suggests that the reaction is of the first order, but the Lindemann theory must be invoked in order to determine whether it is a unimolecular gas reaction.

The two reactions in question 8(a) are quite distinct, that with alkali being commonly referred to as saponification rather than hydrolysis.

The method of studying a reaction by withdrawing samples from time to time and determining a concentration is not applicable to a gas reaction (question 10). The analogue of concentration in the case of a gas is pressure, and changes of pressure (or volume) often provide a convenient means of following a gas reaction. However, such a method will apply only when the reaction produces a change in the total number of moles of gaseous substances present (question 11). Otherwise one must make use of a specific property of the gases involved.

It is rarely possible to find the two rate constants of a reversible reaction directly, that is, by starting at each extreme in turn and finding the rate constants before the reverse reaction becomes appreciable. Generally one rate constant may be found in this way, and the sum of the two rate constants as the reaction approaches equilibrium, provided that the reaction is of the same order in each direction.

In a catalysed reaction (question 14) the rate constant incorporates the concentration of the catalyst, and, where the rate is directly proportional to the catalyst concentration, the rate constant is the product of an intrinsic rate constant and the catalyst concentration, e.g.

$$-\frac{\mathrm{d}x}{\mathrm{d}t} = k[x][\text{catalyst}]$$
$$= k'[x]$$

where $k' = k[\text{catalyst}]$.

The second part of question 19 is merely an application of the Arrhenius theory. A simple potential energy diagram will illustrate the relationship.

Do not attempt a question such as number 21 in an examination unless you have recently carried out such an experiment satisfactorily and are able to remember all the details apart from numerical results. Refer again to Chapter 1.

M. MOLECULAR AND CRYSTAL STRUCTURE

THE structures of molecules and crystals depend on the nature and properties of the bonds involved. If one considers the whole range of possible structures, from the pure ionic crystal to the molecular crystal, it will be appreciated that crystal and molecular structures are, in fact, very closely related. For this reason we have combined our questions on these topics in one section.

All kinds of spectra provide useful physical methods for structural investigations, but, for convenience, we have treated optical and magnetic resonance spectra separately (Section Sp) and X-ray spectra and X-ray diffraction in the present Section.

(i) Definitions

Additive, constitutive, and colligative properties.
Optical exaltation.
Polarizability.
Dielectric constant.
Dipole moment.
Optical rotation.
Specific rotation.
Molecular rotation.
Rotatory dispersion.
Circular dichroism.
The Faraday effect.
Magnetic rotatory power.
Molecular rotatory power.
Magneto-optical exaltation.
Characteristic X-ray lines.
Orientation forces.
Induction forces.
Dispersion forces.

M. MOLECULAR AND CRYSTAL STRUCTURE

(ii) Laws and Equations

The parachor.
Molar refraction equation.
The Mosotti–Clausius equation.
Maxwell's relationship.
The Debye equation for molar polarization.
The Drude equation for rotatory dispersion.
The principle of optical superposition.
Hudson's iso-rotation rules.
Verdet's equation.
Moseley's equation.
Potential energy of attraction:

   Dispersion effect.
   Induction effect.
   Orientation effect.

The Bragg equation.

(iii) Questions

1. Write an essay on the investigation of molecular structure by physical methods. (Ge)

2. Summarise the methods available for the measurement of bond lengths.
   Discuss *one* of the methods you mention in some detail. (Gi)

3. Write briefly on any *two* of the following topics:

   (*a*) The determination of the strengths of chemical bonds.
   (*b*) The different types of chemical bonds in solids.
   (*c*) The nature of intermolecular forces and the factors influencing their magnitude. (Gi)

4. Write an essay on the determination of crystal structure by X-ray diffraction. (R)

5. Show how X-ray diffraction methods have led to a knowledge of the arrangements of the ions and the interionic distances in crystals of alkali-metal halides. (S)

6. *Either:* Give a general account of the evidence that crystalline solids consist of an ordered array of atoms or molecules.
   Or: Describe clearly how the interatomic distances in one particular crystal may be determined. Explain the basic principles involved. (Gi)

7. Explain the difference between a face-centred and a body-centred cubic lattice. Show, with essential experimental detail, how these lattices could be distinguished by means of X-ray diffraction by a powdered sample of a metal. (S)

8. Describe briefly two experimental methods for the determination of the dipole moment of a molecule. Discuss with examples the usefulness in chemistry of a knowledge of dipole moments. (Ge)

9. Answer *either* (a) or (b) in detail *or* (a) and (b) in outline.

(a) Write an account of the determination of the dipole moment of a molecule.

(b) Discuss the use of dipole moments to elucidate molecular structure. (Gi)

10. Describe, with essential experimental detail, *one* method of measuring the dipole moment of a molecule. Discuss briefly the factors which determine the magnitude of the dipole moments of simple molecules.
(S)

11. Write briefly on *two* of the following:

(a) ionisation potentials and electron affinities;
(b) bonding in metals;
(c) directed valency;
(d) dipole moments of molecules. (Gi)

12. Give an elementary derivation of Bragg's Law for the diffraction of X-rays by a crystal. State (without discussing details) what information about a crystal structure is provided by (a) the positions of the lines in the diffraction spectrum, and (b) the intensities of the lines.

Outline the powder method and write notes on the uses and applications of this method. (D)

See also questions E(e). 6, E(f). 3; T. 5; W. 6, 13, 14; X. 20, 24, 31; and Section Sp in general.

Notes. Many more physical methods have been used to investigate molecular structure than can be described in a forty-five-minute essay. Some kind of selection will be essential in order to avoid being too superficial. Essays of this kind so often turn out to be rather boring details of one method after another with few signs of continuity. A fresh approach from a different angle might hold the examiner's attention and gain a better mark. In question 1, for example, try first considering what information on molecular structure is required or desirable, and then proceed to show how differences in these structural factors may affect the physical properties of substances. It should then be possible to discuss the applicability of the various physical methods in some detail. Such an indirect approach is a very useful homework or revision exercise.

The 'strength' of a chemical bond (question 3(a)) is not a clearly defined property but it may be measured in terms of the force constant, dissociation energy, or bond energy. Force constants are calculated from measurements on spectra, dissociation energies either from spectra or calorimetrically, and bond energies calorimetrically. The last are best for comparative purposes since force constants are not simply related to bond strengths and different methods for dissociation energies do not always measure the same quantity. These important differences are often overlooked.

Dispersion forces depend very strongly on the polarizability of molecules (question 3(c)), inductive forces less so, and both inductive and orientation forces depend on the polar nature of the molecules. Hence, the factors which influence the magnitude of these forces are those which govern these two properties of molecules. Brief reference should also be made to the repulsive forces.

In question 4 do not give the impression that the only crystal structure deduced from X-ray diffraction methods is that of sodium chloride. One such simple example is quite inadequate here. Examples should be chosen from different crystal types and different bond types.

The measurement of dipole moments involves two stages (questions 8 to 11). One may determine the dielectric constant at two or more temperatures, or for the solid and the gas (Ebert's method), or the molar refraction at various wavelengths to find $[R]_{Na}$ and $[R]_{\infty}$.

P. PHASE RULE

HERE we shall deal with the phase rule and its applications together with some related topics. For example, questions on Raoult's law and Henry's law appear, whether or not they require actual application of the phase rule, but where there is not a thermodynamic bias. Other questions involving these laws will be found in Section T. Similar comments apply to the distribution law and vapour pressure relationships. Overlapping with the thermodynamics section is unavoidable.

(i) Definitions

Phase.
Number of components in a system.

> If difficulty is experienced in determining the number of components in a system try using the modified phase rule.

Number of degrees of freedom in a system.

> This is not to be confused with degrees of freedom in molecular energy terms.

Triple point.
Metastable state.
Sublimation.
Polymorphism.
Transition point.
Enantiotropy.
Monotropy.
Absorption coefficient (Bunsen).
Coefficient of solubility (Ostwald).
Ideal solution.
Zeotropic mixture.
Azeotropic mixture.
Liquidus and solidus.
Tie-line.
Conjugate solutions.
Critical solution temperature (or consolute temperature).

A system may have an upper or lower critical solution temperature, or both.

Eutectic.

It cannot be too frequently stated that a eutectic is a mixture and *not* a compound. Memorize the evidence for this.

Condensed system.
Congruent and incongruent melting points.
Peritectic.
Eutectoid.
Binodal curve.
Plait point.
Retrograde solubility.

Refer to a triangular diagram with one binodal curve. If one starts with conjugate solutions of a pair of liquids (say A and B) and gradually adds a third (C), the compositions of the (ternary) conjugate solutions will steadily change until they become identical (and so one solution) at the plait point (P). This might lead one to believe that the amount of C at the point P is the *most* that can be involved in a two-phase system, and that any more C would necessarily reduce the system to one phase. But this is not so. A horizontal line drawn between P and the maximum of the binodal curve will show that two phases may be formed in a ternary mixture containing *more* C than at the plait point.

Drying-up point.

(ii) Laws and Equations

The phase rule.

The derivation of the phase rule by thermodynamic methods must be known.

The modified phase rule.

The modified rule proposed by Bowden often gives a clearer result than the original equation. Students are especially advised to use this if they find difficulty in determining the number of components in a system, e.g. in a reactive system or under special conditions such as at a critical point.

Henry's law.
Raoult's law.
The distribution (or partition) law.
Konowaloff's rule.
The expression for the molecular weight of a liquid by steam distillation.

(iii) Questions

1. State Raoult's law and discuss the reasons why departures from this law may arise. Illustrate the effects which such departures may have upon the vapour pressure/composition curve and upon the boiling point/composition curve for binary liquid mixtures. (R)
2. State Henry's law and Raoult's law, showing the relationship between them. Discuss the bearing of deviations from these laws on the separation of liquid mixtures by distillation. (R)
3. Discuss and illustrate the statement that Henry's law and Raoult's law always apply as limiting laws to binary liquid mixtures. (S)
4. Discuss the behaviour which may be encountered during the distillation of (a) mixtures of miscible liquids, and (b) mixtures of immiscible liquids. (R)
5. State and explain Raoult's law. Discuss, with the aid of diagrams, binary systems which show departures from this law and indicate how such departures may arise. (R)
6. State Raoult's law and Henry's law and indicate their significance in relation to the definition of ideal mixtures. Show that these laws apply, as limiting laws, to non-ideal mixtures. Give a brief discussion of how non-ideality arises in various liquid mixtures. (R)
7. Draw sketches to illustrate how the vapour pressure of a mixture of two volatile liquids may vary with composition at constant temperature (both totally miscible and other systems should be considered). State what you mean by an ideal solution, and explain the circum-

stances that may give rise to deviations from the ideal vapour pressure curve. (R)

8. The vapour pressure of water is 4·58 mm at 0° C and 12·68 mm at 15° C.

(a) Calculate the molar latent heat of evaporation of water.

(b) Plot the data on the graph paper provided and hence determine the vapour pressure of supercooled liquid water at −10° C.

(c) Compare the value you obtain with the vapour pressure of ice at −10° C (1·95 mm), and comment. (R)

9. Discuss the distribution of a simple solute between two immiscible solvents in terms of (a) thermodynamic criteria of equilibrium and (b) the Phase Rule.

Describe one example of the use of the distribution law in the study of a chemical process such as complex formation. (R)

10. Give a thermodynamic proof of the Phase Rule. Discuss in terms of the rule the changes which may occur during the isothermal evaporation of an aqueous solution of a salt which forms a series of solid hydrates. (S)

11. Deduce the phase rule $P + F = C + 2$. Discuss with the help of the phase rule and suitable diagrams *either* the equilibrium between liquid and solid phases in systems of two components, *or* the crystallisation of a solid phase by isothermal evaporation of aqueous solutions of a salt which forms several hydrates. (Ge)

12. Write comments on *two* of the following statements:

(a) A two-component system is invariant when four phases coexist.

(b) Salts with a high solubility in water are deliquescent.

(c) Solutions freeze at lower temperatures than the pure solvent. (R)

13. State the phase rule, explaining each of the terms used. The following solutions of manganous nitrate were found, on cooling, to begin to deposit crystals at the temperatures stated:

g $Mn(NO_3)_2$/100 g H_2O	21·3	33·0	40·5	42·3	45·5	50·5
Temperature ° C	−10	−20	−36	−29	−16	0·0
g $Mn(NO_3)_2$/100 g H_2O	54·6	62·4	64·6	65·6	67·4	76·8
Temperature ° C	+11	+25·8	+23·5	+27	+30	+35·5

No liquid phase was observed at any temperature below −36° C. Construct the phase diagram over the composition range involved and interpret it as fully as possible. Label the areas of the diagram clearly to indicate the phases present. (Formula weights: $Mn(NO_3)_2 = 178·96$; $H_2O = 18·016$.) (R)

14. State the *phase rule* explaining the significance of the terms and symbols used. With the aid of this rule describe and interpret *either* (*a*) the melting-point composition diagrams which might be encountered with binary alloys which may form mixed crystals *or* (*b*) the boiling-point diagrams of binary liquid mixtures. (Gi)

15. State the phase rule, explaining each of the terms used. Discuss the equilibrium diagrams for:

either (*a*) a two-component system which shows complete miscibility in the liquid state and in which no compound occurs, but in which solid solutions are formed;

or (*b*) a two-component liquid system in which partial miscibility occurs. (R)

16. Explain what is meant by chemical potential. Use this concept to deduce the phase rule, explaining the meaning of the terms employed. Draw the equilibrium diagrams for a system of two components which are completely miscible in the liquid state and partially miscible in the solid state, (*a*) showing a eutectic point and (*b*) showing a transition point (peritectic point). Discuss these diagrams with the help of the phase rule. (Ge)

17. Derive the *phase rule*, defining the symbols used. Using this rule, discuss the possible phase diagrams of binary systems in which no compounds are formed but in which solid solutions occur. (R)

18. State the phase rule and define carefully all the terms involved. With reference to this rule discuss the behaviour at constant pressure of a two-component system in which no solid solutions are formed and the components are completely miscible in the liquid state, but in which the components form a compound of incongruent melting point. Label clearly any phase diagrams and cooling curves you may draw. (R)

19. Show how the results obtained from an analysis of the cooling curves for a series of mixtures of two metals can be used to verify both the presence of a compound with congruent melting point and its composition. (R)

20. Show by clear diagrams and explanatory notes the behaviour which may be observed on crystallization of two-component systems in which no solid solutions are formed but where compound formation may occur. (R)

21. C and D are two partially miscible liquids. E is a substance that dissolves without chemical reaction in both C and D.

(*a*) By means of diagrams show the possible effect of temperature changes on an equilibrium mixture of C and D.

(*b*) Describe what you will expect to occur when varying amounts of E are shaken with a mixture of C and D.

(*c*) Apply the phase rule to the resulting mixture of C, D and E. (R)

22. Consider the following equilibrium systems in the light of the equation $P + F = C + 2$. Give reasons for the values of C and F which you allot in each case.

(a) liquid water, hydrogen gas and oxygen gas, between 0° C and 100° C, at a total pressure of 1 atmosphere ($P = 2$);

(b) steam, at 1 atmosphere pressure, heated to 2,000° C; dissociation into hydrogen and oxygen appreciable ($P = 1$);

(c) as in (b) but with an equal volume of hydrogen added before heating up ($P = 1$);

(d) solid NH_4HS between 20° C and 25° C, partly dissociated into gaseous ammonia and hydrogen sulphide ($P = 2$);

(e) gaseous H_2S added in excess to gaseous NH_3 at 20° C to 25° C, the pressure being sufficiently high to precipitate some solid NH_4HS ($P = 2$);

(f) a gaseous mixture of hydrogen, carbon dioxide, carbon monoxide, and water at 800° C and a total pressure of 1 atmosphere ($P = 1$);

(g) as in (f) but with the gas phase also in equilibrium with solid carbon ($P = 2$). (R)

23. Explain with the aid of phase diagrams, *three* of the following:

(a) the action of a salt–ice freezing mixture;

(b) the phenomena accompanying the cooling of a homogeneous molten mixture of two metals that form a stable compound, the compound having two interconvertible crystalline forms, and none of the solid phases being mutually soluble;

(c) the isothermal dehydration of a saturated solution of a hydrated salt, such as copper sulphate pentahydrate;

(d) the distillation of a mixture of two liquids that form an azeotrope. (R)

24. Explain what is meant by an ideal solution.

Show, by means of carefully labelled diagrams, how the vapour pressures of mixtures of two volatile liquids (completely or partly miscible) may vary with composition at constant temperature. Suggest possible reasons for the deviations from ideality shown by these systems. (D)

25. State and explain the Gibbs Phase Rule, paying attention to clarity of definition of the terms involved. Discuss *two* types of invariant equilibria which may occur in two-component systems. (C)

26. A liquid (A) is soluble in liquids (B) and (C), which are partially miscible with each other. Draw a diagram to show the ternary conjugate solutions that can exist in equilibrium at a given temperature. Using this diagram, describe what may happen when (A) is added to various two-phase systems containing (B) and (C) alone. (C)

27. Explain, with the aid of phase diagrams, *two* of the following:

(a) The effect of composition on the vapour pressure, at constant temperature, of two miscible liquids deviating from Raoult's Law.

(b) The effect of temperature on the composition of two partially miscible liquids in equilibrium.

(c) The phenomena accompanying the cooling of a homogeneous molten mixture of two components, A and B, that form a compound AB unstable at its melting point, none of the solid phases being mutually soluble. (C)

28. Draw and explain a Phase Rule diagram for a mixture of three liquids, one pair of which is only partially miscible at 25° C. How may the effect of temperature be shown? (C)

29. Discuss the distillation and separation of two liquids which form a constant boiling mixture of minimum boiling point. Mention any uses you know of for an azeotropic mixture. (C)

See also questions T. 25, 26, 27, 31; X. 16, 24.

Notes. Phase rule questions are more liable than most others to bring forth from examination candidates pages of irrelevant information. This suggests that immediately a particular type of system is mentioned students tend to reproduce the whole of their lecture notes on the subject. In question 7, for example, the sketches are to represent the various kinds of vapour pressure–composition curves, but a discussion of the distillation of the mixtures concerned is *not* required. For the second part of this question remember that ideality is more than simply adherence to Raoult's law.

Appropriate diagrams or equilibrium curves, suitably labelled, are generally necessary and convenient in answering phase rule questions, whether or not diagrams are specifically asked for.

Several questions here are very similar, but, as pointed out in Chapter 1, each is designed to emphasize a different aspect of the topic and so calls for a different approach. For a binary liquid system a single diagram, clearly labelled, showing the individual and total vapour pressure curves for both ideal and non-ideal mixtures can be extremely useful. It can indicate the applicability of Raoult's law to the solvent in dilute or ideal solutions and of Henry's law to the solute in dilute or ideal solutions.

This is so important that a suitable diagram is reproduced as Figure 10, in which the broken lines show the *ideal* individual and total vapour pressure curves, and the full lines show the effects of negative deviation from Raoult's law. Such a diagram is particularly appropriate in questions 3 and 6 to show the applicability of the two laws as limiting laws, i.e. at low concentrations. The dotted lines show the Henry's law slopes.

Figure 10

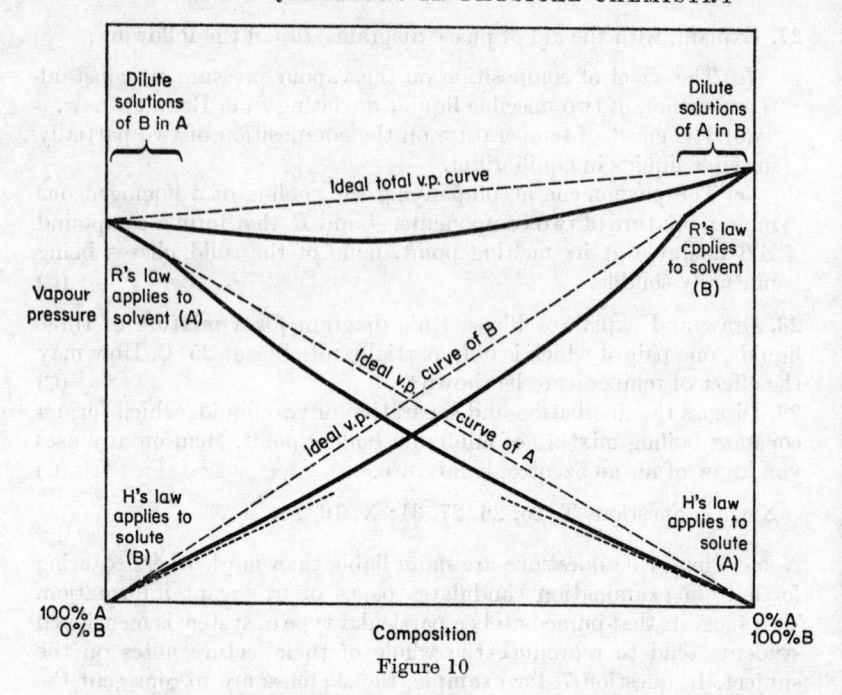

Departures from Raoult's law are best considered as being due to factors which change the *effective* mole fraction of the solvent. Association of solvent molecules, for example, reduces the mole fraction of solute below the calculated value, which increases the mole fraction of solvent and so gives rise to a positive deviation. Compound formation, hydrogen bonding, and similar strong interactions can be shown by a similar argument to cause negative deviations.

Since the normal boiling point of a liquid is the temperature at which its vapour pressure is one atmosphere, then, of two liquids, that with the lower vapour pressure will have the higher boiling point. This means that the boiling-point–composition curve has the shape of the vapour-pressure–composition curve *reversed* and *inverted*.

In the formation of salt hydrates by evaporation of a solution (questions 10 and 11) the phase rule must be applied to the successive equilibria. The latter must be represented by *balanced* equations.

The freezing-point depression method for the molecular weight of a solute is well known, but one must avoid unjustified generalization. The freezing-point method is applicable to non-volatile solutes in very dilute solutions of liquid solvents. A consideration of equilibrium diagrams will show that the statement in question 12(c) requires modification.

It will be seen that many of the questions merely require a familiarity with the commoner types of phase diagram. This can be acquired only by practice in drawing and explaining such diagrams. It is helpful to consider how the more complex diagrams can be broken down into two or more simple forms.

Any convenient concentration units may be used in constructing an equilibrium diagram or phase diagram, as in question 13. Do *not* convert *all* concentration data, for example, to moles per litre; the formula weights are given so that the composition may be calculated in appropriate units for the important points which will appear when the diagram is plotted.

Misunderstanding of the question is a common failing in this subject. Time must be allowed for careful reading and interpretation of questions such as numbers 15, 16, and 18. Jot down rough diagrams as you read if it will help in sorting out the diagrams required.

Questions 22 and 23 are direct applications of the definitions of the phase rule. If in difficulty use the modified phase rule suggested by Bowden.

S. SURFACE CHEMISTRY

ALL phenomena associated with interfaces are included in surface chemistry (with the exception of electrochemical phenomena). The thermodynamics of surfaces should be studied, especially the surface free energy and entropy, and their use in describing adhesion, cohesion, the spreading of liquids, and adsorption of all kinds.

Electrical phenomena at interfaces are very important, particularly in the study of colloidal systems. The nature and formation of the electrical double layer must be understood, including the origins of the zeta potential, essential to an explanation of the various electrokinetic phenomena.

For reasons of convenience courses on surface chemistry are generally divided into separate topics, such as electrokinetic phenomena, adsorption, colloidal systems, and chromatography. As these are all intimately related we shall not sub-divide the questions, although grouping them as conveniently as possible, but the lists of definitions, laws, and equations will be divided.

It will be noted that these questions frequently involve two or more topics which might have been widely separated in the student's lecture course. This emphasizes the danger of trying to keep topics in water-tight compartments.

There are far fewer equations and laws to learn in surface chemistry than in some other subjects, but many different theories. Under 'definitions' will be found some items which the student should be able to explain in a few short sentences, rather than define concisely and precisely.

(i) Definitions

Thermodynamics of Surfaces

Surface free energy and entropy.
Work of cohesion.
Work of adhesion.

> Carefully distinguish these two; only *one* substance is involved in cohesion, but two in adhesion.

Spreading coefficient.
Contact angle.

> This must be measured in the *liquid* phase.

Electrokinetic Phenomena

Zeta potential.
Electrophoresis.

> Originally called cataphoresis. This term is sometimes used for the movement of positively charged *colloidal* particles under a potential gradient, and electrophoresis when *large* solute molecules are involved.

Streaming potential.
Sedimentation potential.

Adsorption

Surface excess.

> At solution–solid interfaces it should be noted that it would be more correct to refer to adsorption *of* solutions rather than adsorption *from* solutions, since both components of the solution are generally adsorbed, even though the surface may show preference for either the solute or the solvent.

Film pressure.
Surface potential.
Isotherm.
Isobar.
Isostere.

> The constant terms in the last three are respectively temperature, pressure, and amount of substance adsorbed.

Accommodation coefficient.

Colloidal Systems

The Hofmeister (or lyotropic) series.
Sensitization of sols.
Protection of sols.
Thixotropy.
Rheopexy.
Dilatancy.
Solubilization.

Chromatography

Height equivalent to a theoretical plate (H.E.T.P.).
R_F value.

Ion Exchange

The capacity of an ion exchanger.

(ii) Laws and Equations

Thermodynamics

The Eötvös equation.
The Ramsay and Shields equation.
> These also have other applications, for example in the investigation of association in liquids.

The Kelvin equation.
The Gibbs adsorption equation,
> i.e. the thermodynamic derivation of the equation.

The Donnan equation (the Donnan membrane equilibrium).

Adsorption

The Langmuir adsorption isotherm.
The classical (or Freundlich) adsorption isotherm.
The Brunauer, Emmett, and Teller equation (B.E.T. equation).
The Harkins–Jura equation.

Colloidal Systems

Einstein's equation for viscosity.
The Weimann equation for rate of condensation.
The Staundinger equation (or rule).

(iii) Questions

1. Discuss the adsorption of gases at the surface of a solid. (R)
2. Give an account of the adsorption of gases by solids, emphasising the quantitative treatment of this subject. (Gi)
3. Give an account of the phenomena associated with adsorption at *either* (a) gas–solid *or* (b) vapour–liquid *or* (c) liquid–solid interfaces. (R)
4. Discuss the phenomena associated with adsorption at *one* of the following interfaces: (a) solid–gas; (b) solid–liquid; (c) liquid–gas. (S)
5. Write an account of any *one*, or *all*, of the following:

    (a) adsorption of solute at a liquid surface;
    (b) adsorption of gases on solids;
    (c) adsorption of solutes from solution at solid surfaces.

(Two or three sections may be answered, but full marks may be obtained for an answer to any one section.) (R)

6. Describe *briefly* the methods for determining adsorption isotherms at gas–solid interfaces. What types of isotherm have been found and what explanations have been advanced to account for them? (R)
7. *Either:* Give an account of the adsorption of gases and vapours by solid surfaces;

    or: Discuss the nature and properties of insoluble films on liquid surfaces, describing briefly the main experimental methods, used for their study. (Gi)

8. Discuss, with reference to gas–metal systems, the criteria which may be used to distinguish between physical and chemical adsorption. (S)
9. Write an essay on soluble and insoluble monolayers at the surface of water. (Ge)
10. Explain the Gibbs adsorption equation and show its importance in the consideration of films of surface-active substances at a water surface. (Ge)
11. Explain the concept of an electrical double layer existing at the boundary between a liquid and a solid, and describe the use of this concept in the discussion of electrokinetic effects and zeta-potential. How can the zeta-potential be determined? (Ge)
12. Write a short essay entitled 'The colloidal state of matter'. (R)
13. Describe and discuss the properties of lyophobic colloids. (R)
14. Write an essay on the preparation and properties of lyophobic colloids. (R)
15. Describe the evidence for the view that colloidal solutions contain particles of greater than normal molecular size. (R)
16. Write informative notes on *three* of the following:

    (a) the Tyndall effect;
    (b) the Schulze–Hardy rule;

S. SURFACE CHEMISTRY

    (c) the isoelectric point;
    (d) ultra-filtration;
    (e) electrophoresis (cataphoresis). (R)

17. Give a general account of the properties of colloidal electrolytes. (Ge)

18. Write an essay on *either*
    (a) colloidal electrolytes; *or*
    (b) the stability of lyophobic colloids. (Ge)

19. Write an essay on *either*
    (a) the colloidal state; *or*
    (b) adsorption at gas–solid interfaces. (R)

20. Write briefly on *two* of the following:
    (a) colloidal electrolytes,
    (b) electrophoresis,
    (c) the Donnan equilibrium,
    (d) ion exchange resins. (Gi)

21. Discuss *two* of the following:
    (a) the Donnan membrane equilibrium;
    (b) the Gibbs adsorption equation;
    (c) the electrical double layer at an interface between a solid and an electrolyte solution. (Ge)

22. Describe one method of determining the surface tension of a liquid. Dilute aqueous solutions of butyric acid have a much lower surface tension than has pure water. Explain in general terms the cause of this effect, and mention briefly any other experimental evidence that supports your explanation. (R)

23. Write a short essay on *either*
    (a) Adsorption; *or*
    (b) The nature of acids and bases. (R)

24. *Either* write an essay on *one* of the following topics, *or* give a brief account of *two* of the following topics, in either case giving particular reference to analytical applications:
    (a) adsorption chromatography;
    (b) partition chromatography;
    (c) gas–liquid chromatography;
    (d) polarography. (Ge)

25. Write a brief essay on *one* of the following topics:
    (a) The stability of colloidal solutions;
    (b) Electrokinetic phenomena;

(c) The mechanism and kinetics of the synthesis of some common polymer. (Gi)

26. Electrostatic forces are important in determining the properties and behaviour of (a) solutions of strong electrolytes and (b) hydrophobic colloids. *Either* discuss the whole of this statement in general terms, *or* discuss one part of it in some detail. (R)

27. Write an essay on *one* of the following:

(a) specific heats of solids;
(b) colloidal solutions;
(c) heterogeneous catalysis. (Gi)

28. *Either*

(a) Explain clearly the Gibbs Adsorption Equation and show how it has been experimentally verified;

or

(b) Write an account of surface films of insoluble substances. (D)

29. Describe the formation of an electrical double layer at the interface between a solid and an aqueous solution. Show particularly the nature of the zeta potential. Discuss the importance of the zeta potential to the stability of lyophobic sols. (D)

See also questions E(e). 3; X. 25, 26, 27, 29, 32.

Notes. It will be seen that questions on surface chemistry are generally of a descriptive nature, mainly non-mathematical, and providing many opportunities for the practice of essay writing and application of the precepts outlined in Chapter 1. Avoid particularly the reproduction of lecture notes.

The questions listed here contain some typical examples of unnecessary complexity in the instructions to candidates, e.g. in questions 5 and 24. Question 5 asks first for an account of one or all (i.e. three) of the topics listed, but a note in parentheses indicates that either one, two, or three of the sections may be answered. Instructions such as these need very careful consideration, as they may easily cause confusion under examination conditions.

Errors sometimes creep into examination questions; a minor example is the unwanted comma after 'methods' in question 7.

Note the important differences between questions 1 and 2. The latter calls for more detail concerning the methods of measurement of gas adsorption, the thermodynamic treatment, and the derivation of the more important adsorption isotherms. The applicability of these should also be discussed. This is a forty-five-minute question, whereas question 1 is allocated only thirty-six minutes.

An account of the adsorption of solutes at solution surfaces must be mainly concerned with the application of the Gibbs adsorption equation. The meaning of the 'surface excess' must be fully explained. This can be done only by referring the concentrations in the surface and bulk phases to the *same* reference concentration or quantity. For example, if the surface phase concentration is measured as grammes of solute to grammes of solvent, then this must be compared with the concentration of solute in the quantity of the bulk phase which would have the *same* total weight. It is a question, literally, of knowing where to draw the line.

The surface excess increases with concentration to a limiting value corresponding to a complete monolayer. This kind of adsorption may be studied by means of the 'PLAWM' trough.

Adsorption of gases on solids has probably received more attention than other forms of adsorption, possibly owing to its applications, particularly in the determination of the surface areas of adsorbants. An account might be written (e.g. question 5) so as to emphasize the differences between physical adsorption and chemisorption, or with the emphasis on the various kinds of theoretical isotherm, the empirical equation of Freundlich, the kinetic approach of Langmuir and of Brunauer, Emmett, and Teller, and the application of equations of state.

There are two ways of approaching the question of the adsorption of solutes from solution at solid surfaces. One may regard the effect as a distribution between bulk and surface phases, or, by ignoring the solvent, as an adsorption phenomenon analogous to that of gases on solids. In using the latter method discuss the applicability of the common adsorption isotherms.

Sp. SPECTRA

THE lack of questions on spectra in Grad. R.I.C., Part I, papers arises from the fact that the subject was introduced into the syllabus only from 1966. Prior to that date, when no syllabus was published, spectra were clearly regarded as Part II material. We can expect fairly frequent questions on spectra in future in Part I of the Grad. R.I.C. examinations.

(i) Definitions

Free and forced vibrations.
Damped vibrations.
Rayleigh scattering.

> This relates to non-homogeneous systems and is of importance in colloid chemistry. It is included here since it is a result of the interaction of light and matter.

Resonance radiation.
Optical dispersion.

Spectral term.
Rydberg constant.
Multiplicity of a spectral term.
Russell–Saunders coupling.
j–j coupling.

These two forms of coupling angular momenta are extreme cases. The coupling in actual atoms is intermediate between these extremes.

Flame or arc spectrum.
Spark spectrum.
Critical potential,
  which includes:
  ionization potential and
  resonance potential.
Force constant.
Pre-dissociation spectrum.
Induced pre-dissociation.
Zeeman effect.
Nuclear quadrupole.
Gyromagnetic ratio.
Intensity of radiation (or radiant power).
Optical density (or extinction, or absorbance).
Extinction coefficient (or absorptivity).
Transmission (or transmittance).
Molar extinction coefficient (or molar absorptivity).

(ii) Laws and Equations

The Balmer equation.
The Ritz combination principle.
The alternation law.
The Heisenberg uncertainty principle.
Moseley's equation for X-ray spectra.
Beer's law.
Lambert's law.
The spectroscopic displacement law.
Equations for the change in rotational and vibrational energy.
The Franck–Condon principle.
The Morse equation.

(iii) Questions

1. Derive the Beer–Lambert Law for light-absorption and suggest reasons why there may be deviations from it. How would you make a quantitative analysis using u.v. spectra of a solution of two known compounds whose u.v. absorption spectra fall completely within the same wavelength range? (R)

2. Discuss the techniques involved in the applications of *either* (*a*) ultra-violet, *or* (*b*) infrared spectroscopy. (R)

3. Explain briefly the chief features of potential energy curves of diatomic molecules in their ground states and in excited states. Discuss the importance of such curves in interpreting spectroscopic observations. (Ge)

4. Write an essay on the application of spectroscopic measurements to investigations of molecular structure. Limit your discussion to the infrared, visible and ultraviolet portions of the spectrum. (Ge)

5. Discuss the significance of the four quantum numbers used to define the states of the hydrogen atom. Explain briefly why the spectra of other atoms are more complex than the spectrum of atomic hydrogen. (S)

6. Outline the main features of the spectrum of a vapour consisting of atoms of an alkali metal. Discuss the interpretation of such a spectrum. (Ge)

7. Show how the interpretation of the spectra of alkali metal atoms, in the presence and in the absence of magnetic fields, has led to information about atomic quantum numbers. (S)

8. The angular momentum of electrons in atoms may give rise to magnetic moments. Discuss the evidence for, and the interpretation of, such magnetic moments in alkali metal atoms in their various energy levels. (S)

9. Show how the moment of inertia, fundamental vibration frequency, and the anharmonicity constant of a diatomic molecule may be determined from spectroscopic measurements. Indicate the use of these quantities in the calculation of thermodynamic properties. (S)

10. Outline the theory of the rotation spectra of diatomic molecules. (S)

11. Explain the significance, for the spectroscopy of diatomic molecules, of any *three* of the following: (*a*) force constant, (*b*) Franck–Condon principle, (*c*) Morse curve, (*d*) dipole moment and polarisability. (S)

12. Discuss the use of *three* of the following in the investigation of chemical problems:

   (*a*) nuclear magnetic resonance;
   (*b*) Raman spectra;
   (*c*) microwave spectra;
   (*d*) fluorescence;
   (*e*) electron spin resonance. (Gi)

13. Explain, as quantitatively as possible, the following features of spectra:

(a) the infrared spectrum of HCl differs from that of DCl;

(b) H₂ shows a Raman but not an infrared spectrum whereas HCl shows both Raman and infrared spectra;

(c) the ultraviolet absorption spectra of atomic vapours consist of sharp lines whereas those of molecules in solution are broad bands.

(S)

14. Explain the nature of the infrared vibration rotation bands of a diatomic molecule, showing how the measurement of the frequencies of the lines of such a band enables internuclear distances to be determined. (Gi)

15. Molecules of the diatomic gas XY absorb infrared radiation near 1,000 cm.$^{-1}$. At higher resolution this band is seen to consist of a series of approximately equally spaced maxima, 10 cm.$^{-1}$ apart. Discuss the inferences which can be made from these data concerning the properties of the molecule. (Gi)

16. The electronic absorption spectrum of a diatomic molecule may be continuous or may consist of either sharp bands or diffuse bands. Explain, with the aid of diagrams, the significance of each of these kinds of absorption and indicate the information about the molecule that may be deduced from them. (Gi)

17. State, indicating their significance, the laws of photochemical importance associated with the names: Grotthus–Draper, Beer–Lambert, Einstein.

Explain, referring to examples, the occurrence of photochemical reactions with various quantum yields. (D)

See also questions E (*f*). 8; M. 6; X. 27, 31, 32.

Notes. Deviations from the Beer–Lambert law may be the result of physical, chemical, or instrumental factors. Indicate how these effects may be removed or reduced in order to make quantitative spectrophotometric analysis possible. This should be emphasized in the second part of question 1, where the method depends on the strict application of the law and on the additivity of absorption intensities.

Incidentally, there is considerable laxity in the naming of instruments here. A spectroscope is used for *viewing* spectra, a spectrometer may be used for measuring appropriate wavelengths or frequencies, and a spectrophotometer is designed to measure the intensities of lines or bands of spectra.

Potential energy curves are invaluable in helping to explain many phenomena associated with spectra, but they are inevitably limited to diatomic molecules and even here can only be approximate. Do not neglect the correct relative positioning of the curves; for each higher electronic energy level the minimum of the curve is at a greater internuclear distance. This follows logically from the fact that excitation of electrons weakens the bond.

Questions 6, 7, and 8 relate to alkali metal vapour spectra. The emphasis in question 8 is on the Zeeman effect, but the general nature of the spectra will need to be described before the Zeeman effect can be interpreted.

The much more complex spectra of diatomic molecules must be explained in answer to question 9. The last part of this question is, perhaps, rather advanced for Part I, but will well repay investigation if only to emphasize the unity of the molecular and thermodynamic approaches.

Assuming that question 10 relates to pure rotation spectra it presents an opportunity for a clear and concise exposition of the origin, nature, and usefulness of the simplest of all molecular spectra, e.g. the simple selection rule and equal spacing of lines. Only *polar* diatomic molecules will give this kind of spectrum, from which bond-lengths may be determined.

The effects of polarity and polarizability of molecules appear again in question 11(d). In addition to the effect just mentioned there is the very important question of the complementary nature of infra-red and Raman spectra of diatomic molecules. Sections (b) and (c) give further opportunities for the use of potential energy curves.

Nuclear magnetic resonance (N.M.R.) and electron spin resonance (E.S.R.) spectra have not yet become regular Part I topics. The theory of these methods properly belongs to the Part II course, but chemical applications might be asked for in Part I papers, as, for example, in question 12.

The only effective difference between HCl and DCl is in the masses of the hydrogen and deuterium atoms. This will affect the reduced masses of the molecules and hence the moments of inertia. Such a logical approach could produce a good answer to the question (13) even if the isotope effect had not been met previously. The difficulties of far infrared spectrophotometry and the high resolution required will indicate that the isotope effects are too small to be detected in pure rotation spectra.

Potential energy curves are required again in the last-but-one question. With appropriate examples they can be used to explain the origins of line spectra, continuous spectra, and pre-dissociation spectra.

T. THERMODYNAMICS

THERMODYNAMICS is a method or tool rather than a subject on its own account. A knowledge of the theory of chemical thermodynamics without some experience in its application to chemical problems is of very little value. It should be regarded as an alternative to the kinetic approach, to which it is in many ways complementary.

As far as possible we shall arrange the sections in the following order:

>General thermochemistry
>Heat capacities
>Chemical equilibrium
>Vapour pressure relationships
>Miscellaneous topics

(i) Definitions

Specific heat.
Heat capacity.
Molar heat capacity.

>Far too often these three terms are used indiscriminately as though they were identical. Specific heat refers to *one gramme* of a body or substance, and molar heat capacity clearly relates to one mole of a substance, whereas heat capacity is defined without reference to quantity. In physical chemistry the molar heat capacity is the most useful of the three but it is becoming common practice to call it merely the heat capacity.

Maximum work of a process.
Standard enthalpy.
Standard free energy.
Standard entropy.

>The last three also require definitions of the standard states commonly used for all states of matter.

Inversion temperature.
Extensive property.
Intensive property.
Fugacity.
Activity.
Partial molar quantity.
Chemical potential.

(ii) Laws and Equations

The first law of thermodynamics.
Hess's law of constant heat summation.

T. THERMODYNAMICS

Kirchhoff's law.
The second law of thermodynamics.

There are many ways of expressing this law. It is suggested that at least two *distinct* statements should be memorized.

Le Chatelier's principle.
Henry's law.
Raoult's law.
The distribution law.

The simple relationships between the last three laws must be known, and the applicability of Henry's and Raoult's laws in the various concentration ranges. These are conveniently represented on a graph of partial and total vapour pressure against composition.

Carnot's theorem.
The third law of thermodynamics.
The Nernst heat theorem.

In thermodynamics, numbers of equations have to be remembered, only a few of which are known by name. It will be necessary here to list the actual equations, with names where applicable. (*See* appendix for list of symbols.)

$$\Delta U = q - w$$

$$\left(\frac{\partial U}{\partial T}\right)_v = C_v \quad \text{and} \quad \left(\frac{\partial H}{\partial T}\right)_p = C_p$$

$$\left(\frac{\partial \Delta U}{\partial T}\right)_v = \Delta C_v \text{ and } \left(\frac{\partial \Delta H}{\partial T}\right)_p = \Delta C_p \text{ (Kirchhoff's equations)}$$

$$w = nRT \ln(V_2/V_1) \text{ (for gases)}$$

$$w = \int P dV$$

$$C_p - C_v = R \text{ (for an ideal gas)}$$

$$P_1 V_1^\gamma = P_2 V_2^\gamma, \text{ where } \gamma = \frac{C_p}{C_v}$$

$$dS = \frac{\delta q_{\text{rev.}}}{T}$$

$$dU = TdS - PdV$$

$$\left(\frac{\partial U}{\partial S}\right)_v = T \quad \left(\frac{\partial G}{\partial P}\right)_T = V$$

$$\left(\frac{\partial S}{\partial V}\right)_U = \frac{P}{T} \quad \left(\frac{\partial G}{\partial T}\right)_p = -S$$

$\Delta S = nR \ln(V_2/V_1)$ (expansion of an ideal gas)

$H = U + PV$
$F = U - TS$
$G = H - TS$ represented by
$G = F + PV$

$$\begin{array}{ccc} U & \underline{\quad +PV \quad} & H \\ \Big\downarrow -TS & & \Big\downarrow -TS \\ F & \overline{\quad +PV \quad} & G \end{array}$$

The internal energy is often represented by E; if this symbol were used here we should have the alphabetical arrangement $EFGH$ in an anticlockwise direction, with PV added on going from left to right and TS subtracted on going downwards.

For appropriate changes in the various thermodynamic functions we have:

$\Delta H = \Delta U + P\Delta V$

$\Delta F = \Delta U - T\Delta S$

$\Delta G = \Delta H - T\Delta S$

$\Delta G = \Delta F + P\Delta V$

$\Delta S = C \ln(T_2/T_1)$ or $\int \dfrac{C}{T} \mathrm{d}T$ or $\int C \, \mathrm{d}\ln T$

$\dfrac{\mathrm{d}P}{\mathrm{d}T} = \dfrac{\Delta H}{T\Delta V}$ (the Clapeyron equation)

$\dfrac{\mathrm{d}\ln P}{\mathrm{d}T} = \dfrac{\Delta H}{RT^2}$ (the Clausius–Clapeyron equation)

$\dfrac{\mathrm{d}T}{\mathrm{d}P} = \dfrac{T(v_1 - v_s)}{\Delta H}$

$\ln(P_1/P_2) = -\dfrac{\Delta H}{R}\left(\dfrac{1}{T_1} - \dfrac{1}{T_2}\right)$
} forms of the Clapeyron equation.

$\Delta G = RT \ln(p_2/p_1)$

$\sum n_i \overline{\mathrm{d}G_i} - V\mathrm{d}P + S\mathrm{d}T = 0$ (the Gibbs–Duhem equation)

$\Delta G^\circ = -TR \ln K$ (the van't Hoff isotherm)

$\Delta G = \Delta G^\circ + RT \ln Q_a$ (the reaction isotherm)

$\dfrac{\mathrm{d}\ln K}{\mathrm{d}T} = \dfrac{\Delta H^\circ}{RT^2}$ (the van't Hoff isochore)

(iii) Questions

1. State and explain Hess's Law of Constant Heat Summation, and discuss its relationship to the First Law of Thermodynamics.

The amounts of heat evolved during the combustion of 1 mole of gaseous ethane, gaseous hydrogen, and graphite are 373, 94, and 136 kcal, respectively, at 25° C and 1 atm pressure. Calculate the heat of formation of ethane from graphite and molecular hydrogen at 1 atm pressure. What would be the difference between the heats of formation at constant pressure and constant volume? (R)

2. The heat of combustion at constant volume of methyl alcohol to carbon dioxide and liquid water at 25° C is 173·34 kcal mole^{-1}.

Describe briefly but with essential details an experimental method by which results such as this may be obtained. Calculate the heat of the reaction at constant pressure, and comment on the value of heats of combustion in physical chemistry. (R)

3. Define 'intensive property' and 'extensive property'. Review the calorimetric and non-calorimetric methods which are available for the determination of heats of reaction at constant pressure over a range of temperatures. Choose *one* of the non-calorimetric methods you mention for more detailed discussion, including (a) the derivation of formulae, and (b) illustration by means of a specific example. (Ge)

4. State the fundamental laws of thermochemistry and demonstrate that they arise from thermodynamic principles. Discuss methods of determining heats of formation of chemical compounds and the implications that can be drawn from the results. (S)

5. State Hess's Law of constant heat summation and indicate how it is related to the first law of thermodynamics.

Show how Hess's law may be applied to the evaluation of *two* of the following:

(a) lattice energies of crystals;
(b) heats of solvation of electrolytes;
(c) the bond energy of the C–H bond in methane (mean bond energy). (S)

6. Give an account of the principles of methods which may be used to find the change in enthalpy (ΔH) in a chemical reaction. (Ge)

7. Discuss the importance of heat capacity determinations in physical chemistry. (R)

8. The molar heat capacity at constant pressure of nitrogen is 6·93 cal deg^{-1} at 15° C, and increases with increasing temperature; that of argon is 5·00 and is independent of temperature. Explain these facts as fully as possible.

How are heat capacities useful in thermochemical calculations? (R)

9. Give precise and general definitions of 'heat capacity'. Determinations of heat capacity have played, and still play, an important part in pure and applied thermodynamics. Write a concise essay showing how and why. (R)

10. Discuss the molar heat capacities of a gas at constant volume and at constant pressure, and the difference between them. Show how the

temperature variation of the heat change attending a reaction can be deduced from heat capacity data. (R)

11. Define the heat capacities of gas at constant pressure (C_p) and at constant volume (C_v). Show how these differ for (*a*) a real gas, and (*b*) an ideal gas. The measured values of C_v for argon, nitrogen and carbon dioxide, all at 25° C, are 2·98, 4·93 and 6·75 cal deg⁻¹, respectively. Discuss these values with reference to the principle of equipartition of energy. (R)

12. Explain the meanings of the terms *chemical equilibrium* and *equilibrium constant*, and derive an expression relating the latter to the change in free energy occurring in the reaction.

The equilibrium constant K_p for the reaction $PCl_5 = PCl_3 + Cl_2$ is 0·31 atm. at 200° C and 3·24 atm. at 274° C. Calculate the standard free energy change of the reaction at 200° C and the mean change in the heat content of the reaction over the temperature range 200° C–274° C. (R)

13. Discuss thermodynamic criteria for chemical equilibrium:

(*a*) at constant volume and temperature, and
(*b*) at constant pressure and temperature.

Using the concept of chemical potential derive the formula

$$-\Delta G° = RT \ln K_p$$

and discuss its application. (Gi)

14. Discuss the criteria of equilibrium based on the second law of thermodynamics. Define all the quantities you mention and show how the criteria are interrelated. Discuss in the widest thermodynamic terms the reaction $H_2 + \tfrac{1}{2}O_2 = H_2O$ occurring over a wide range of temperature. (S)

15. State a thermodynamic criterion for chemical equilibrium and indicate its theoretical basis. Derive a relationship between the equilibrium constant of a chemical reaction and the free energy change of the reaction.

At 300° K and 1 atm pressure, dinitrogen tetroxide is 50 per cent dissociated into nitrogen dioxide. Calculate the equilibrium constant and standard free energy change for the reaction $N_2O_4 = 2NO_2$ at this temperature. (R)

16. Starting from the relation between the equilibrium constant (K_p) and the Gibbs free energy change (ΔG) for a gaseous reaction, derive an expression for the variation of K_p with temperature. Describe briefly how the equilibrium constant of a reaction of your own choice could be measured.

Given that $K_p = 0·01$ atmospheres⁻¹ at 27° C and $\Delta H = +15,000$ cals. mole.⁻¹ (independent of temperature), calculate K_p at 127° C. ($R = 1·99$ cal. degree⁻¹mole⁻¹.) (Gi)

17. Derive the thermodynamic relationship between the change in free energy associated with a chemical reaction and its equilibrium constant.

The equilibrium constant K_p for the dissociation of dinitrogen tetroxide into nitrogen dioxide is 1·34 atm at 60·2° C and 6·64 atm at 100·1° C. Determine the free energy change of this reaction at each temperature and the mean heat content (enthalpy) change over the temperature range. What inferences can be made about the change in standard entropy of the reaction? (R)

18. Derive a differential equation which expresses the variation with temperature of the equilibrium constant, K_p, of the hypothetical gas phase reaction,

$$2A = B$$

Explain how the value of such an equilibrium constant at one temperature may be calculated from its value at a *substantially* different temperature. What experimental data are required for the calculation? (S)

19. Explain what is meant by the term chemical potential and discuss its application to one chemical problem. The equilibrium constant for the dissociation of hydrogen iodide is 0·02515 at 508° C and 0·01679 at 393° C. Calculate the mean standard entropy change of the reaction over this temperature range. (R)

20. What is meant by chemical potential? Using this concept derive a relation between the equilibrium constant of a homogeneous chemical reaction and the free energy change in the reaction. Show, with reference to some technical process, how a knowledge of the dependence of the equilibrium constant on pressure and temperature is of importance in deciding upon the best operating conditions. (Gi)

21. Thallous hydroxide dissociates according to the equation: $2TlOH = Tl_2O + H_2O$. At 100° C the dissociation pressure of the system is 125 mm. Calculate the free energy change of the reaction.

What further information would you require to calculate the entropy change? Would you expect this to be positive or negative? (R)

22. State the conditions of equilibrium for a system at constant temperature and pressure. Hence derive a relationship between the standard free energy change of a homogeneous gas reaction and the equilibrium constant of the reaction. Give a brief account of the importance of standard free energy changes. (Ge)

23. For the reaction: $2CO_2 = 2CO + O_2$, $K_p = 4·00 \times 10^{-21}$ at 1000° K and $K_p = 1·03 \times 10^{-12}$ at 1400° K.

Use these data to calculate (a) ΔH for the reaction, (b) the partial pressure of oxygen in the equilibrium mixture at 1000° K and one atmosphere pressure. (R)

24. From the fundamental equations of thermodynamics derive the relationship between the temperature variation of the pressure at which two phases are in equilibrium and the latent heat of the change of phase. Show how the expression may be modified when one of the phases behaves effectively as an ideal gas.

The densities of ice and water at 0° C and 1 atm pressure are 0·918 and 1·000 g cm^{-3}, respectively. The latent heat of fusion of ice is 79·67 cal g^{-1}. Estimate the melting point of ice under a pressure of 200 atm if the densities remain practically constant over this pressure and temperature range. (R)

25. Derive a relationship expressing the variation of the vapour pressure of a liquid with temperature. State clearly the approximations made.

The vapour pressure of carbon tetrachloride at various temperatures is as follows:

Temperature (° C)	19·2	41·2	60·8	76·4
Vapour pressure (mm)	88·0	222·8	452·8	760·0

By graphical means or otherwise interpret these results on the basis of the relationship you have derived. (R)

26. The following values have been reported for the saturated vapour pressure of mercury at the temperatures stated:

t (° C)	50	80	100	150	200
p (mm Hg)	0·0122	0·0885	0·276	2·88	17·81

Calculate the mean latent heat of vaporization and also the entropy of vaporization at the normal boiling point of mercury. Derive any equation you use. (R)

27. Describe a method by which the vapour pressure of a pure liquid may be determined over a range of temperatures. The vapour pressure of diethyl ether is 185 mm at 0° C, and 921 mm at 40° C. Calculate the latent heat of vaporization, assumed constant over this temperature range. At what temperature would the ether boil at 760-mm pressure? (R)

28. Measurements of the vapour pressure of a pure liquid over a range of temperatures are frequently expressed by means of an equation of the form $\log P = A - B/T$, where A and B are constants and T is temperature in ° K. Show from thermodynamic principles what is the fundamental basis of this equation, and discuss the approximations which it may involve. Describe briefly *one* method by which such measurements can be made. (Gi)

29. State the thermodynamic condition for equilibrium between two phases in a one-component system, and from this derive a relationship between the vapour pressure of a liquid and its temperature. Describe how the variation of the vapour pressure of a pure liquid with temperature can be studied experimentally. (R)

30. What is meant by the free energy change (ΔG) of a reaction? How does it differ from the enthalpy change (ΔH) of the reaction?

Describe one method by which the standard free energy change of a reaction can be obtained. (R)

31. Deduce, from the concept of chemical potential, the van't Hoff isotherm for the change in Gibbs free energy associated with a chemical reaction at constant temperature. Discuss on this basis the effect of changes in (a) the composition of the gas phase and (b) the total pressure of the gas phase on the position of equilibrium in the reaction:

$$NiO(s) + H_2(g) = Ni(s) + H_2O(g).$$

Show that your conclusions are consistent with the Phase Rule. (S)

32. Outline the part played by entropy in the thermodynamic treatment of equilibria and discuss the statistical interpretation of entropy. (Ge)

33. Define the symbols used in the thermodynamic equations: $G = U + pV - TS$ and $H = U + pV$.

Show how using these relations and the condition that for an infinitesimal reversible process $dU = TdS - pdV$, it is possible to obtain a relationship between ΔH, ΔG, and $d\Delta G/dT$ for a chemical reaction. Discuss the applications of this relationship. (R)

34. Discuss, in terms of the first and second laws of thermodynamics, the principles of methods which are available for the determination of changes in enthalpy (ΔH) and free energy (ΔG) in chemical reactions. (Ge)

35. Discuss the nature of entropy, and briefly outline how the entropy of a diatomic gas may be determined. (Gi)

36. What is meant by entropy? Discuss this concept in terms of the disorder of the system with particular reference to (a) the entropy change at a transition point, (b) the entropy change when a perfect gas expands at constant temperature.

Indicate how the standard entropy of a pure chemical substance may be determined. (Gi)

37. Extensive tabulations of the standard Gibbs free energies of formation ($-\Delta G°$) of chemical compounds are available. Enumerate the methods by which these standard data are obtained, showing in each case the fundamental principles involved. (Gi)

38. The transfer of heat from a hot body to a cold body is a natural process which can be adapted to the purpose of providing useful work. Discuss this statement, and show by means of a reasoned argument what is the maximum efficiency of a cyclic process designed to perform this task. Comment on any fundamental implications of the result you obtain. (R)

39. Give an account of the critical state, and show how the temperature, pressure and volume of a substance at the critical point may be determined. (R)

40. Write briefly on *three* of the following:

(*a*) Entropy of mixing.
(*b*) The Joule–Thomson effect.
(*c*) Heats of formation of compounds.
(*d*) Trouton's rule. (R)

41. State Raoult's law and Henry's law, and explain their importance in the thermodynamic treatment of ideal and real liquid systems in terms of activities. (S)

42. Give a brief account of the experiments performed by Joule and Thomson on the expansion of gases through a porous plug and discuss the thermodynamic theory of the effect. Explain why the Joule–Thomson experiment may either lower or raise the temperature of the gas involved, and comment briefly on the significance of the experiment for the liquefaction of gases. (S)

43. Assuming Raoult's law, derive an expression for the effect of a solute on the freezing point of a dilute solution. Discuss, in terms of your expression, the freezing points of solutions of (*a*) neutral molecules in non-polar solvents and (*b*) salts in polar solvents. (S)

44. The chemical potential of a substance in solution may be represented by the equation

$$\mu_i = \mu_i^\circ + RT \ln m + RT \ln \gamma$$

and has its standard value in the standard state defined by $m = 1$ and $\gamma = 1$. Define all the terms in this equation and explain its significance. Briefly discuss the factors which influence the value of γ in the case of a solute which is a strong electrolyte. (Ge)

45. Calculate the entropy change, for 1 mole of an ideal gas, for each of the following processes:

(*a*) the reversible isothermal compression of the gas at 300° K from a volume of 20 litres to 5 litres;

(*b*) the adiabatic expansion of the gas, initially at 300° K, from a volume of 5 litres to 20 litres;

(*c*) the irreversibly performed expansion of the gas from a volume of 5 litres at 300° K to 20 litres at 300° K.

($R = 1.99$ cal.deg.$^{-1}$ mole^{-1}; $\ln 10 = 2.30$.)

Outline the theory underlying your calculations, and state whether any of the data provided for these calculations are redundant, or whether insufficient information is given. (S)

46. Give a definition of 'partial molar property'. State reasons why the partial molar properties of a component in a mixture may be the same as, or different from, the corresponding molar properties of the same component in the pure state. Derive the Gibbs–Duhem equation and discuss one of its applications to two-component systems. (Gi)

47. Assuming the Maxwell–Boltzmann distribution law, derive a statistical expression for the entropy of a monatomic gas. How may this expression be tested experimentally? (S)

48. The entropies of the perfect crystalline forms of all pure substances at $0°$ K are by convention taken to be zero.

Discuss this statement from the point of view of statistical mechanics.

Explain the relevance of the convention to the determination of entropies, and indicate how the entropies of pure substances in the solid, liquid and gaseous states are determined by calorimetric measurements. (S)

49. A 200-ml container is filled at $15°$ C. with $4 \cdot 00 \times 10^{-3}$ mole of the ideal gas A. What is the pressure of A? The temperature is raised to $400°$ C., and under these conditions A reacts reversibly according to the equation

$$A \rightleftharpoons B + C \quad (\Delta H = 10 \text{ kcal.})$$

with a dissociation equilibrium constant (K_p) of $3 \cdot 10$ atm. If B and C are gases also displaying ideal behaviour, what is the partial pressure of A at equilibrium at $400°$ C., assuming the volume of the container to remain constant? Calculate also the value of K_p at $450°$ C., and comment briefly upon your calculations. (Ge)

50. You are required to determine the Gibbs Free Energy change, ΔG, which accompanies the formation of the hydrate $CuSO_4.5H_2O$, from anhydrous $CuSO_4$ and liquid water. What data would you require, and how would you proceed with the calculation? Indicate clearly the principles of the method you use. (D)

51. Define 'Specific Heat' and 'Heat Capacity'. Discuss the importance of heat capacity determination in chemical thermodynamics. (D)

52. What is 'Thermodynamic Reversibility'? Discuss reversibility in relation to the Second Law of thermodynamics. (C)

53. Derive the Clausius–Clapeyron equation, which relates pressure, temperature and volume in a physical change.

From this equation and Raoult's Law derive an expression for the elevation of the boiling point of a dilute solution of a non-volatile solute.

54. A gas is initially in a state T_1, P_1 and is changed to a state T_2, P_2. Devise two reversible paths between these states and show that the heat absorbed by the gas is not the same along the two paths, but that the entropy change is the same.

Assume that the gas is perfect and that its heat capacity is constant. (C)

55. For a chemical reaction how are the following related to one another: the maximum work; the net work; the Work Function; the Gibbs Free Energy?

At 102° C. and a total pressure of 1 atmosphere, sulphuryl chloride is dissociated according to the equation

$$SO_2Cl_2 = Cl_2 + SO_2$$

to the extent of 91·2%. Find K_p and the net work of the reaction at this temperature. (C)

56. What is an 'energy level diagram' and on what thermodynamic principle are such diagrams based? From the information given below construct energy level diagrams for the reactions of tin and of copper with dilute acids. Show how the diagrams indicate the causes of the different reactivities of the two metals.

	Tin	*Copper*
Standard heat of formation of X^+ (g)	72	81·5 kcal/mole
Standard heat of formation of X^{++} (g)	581	730 ,, ,,
Hydration energy of X^{++}	451·4	592·1 ,, ,,

$$2H^+(aq) = H_2(g): \Delta H° = -204 \text{ kcal/mole}$$

(C)

57. Write short notes on *three* of the following:

(a) Kirchhoff's equations.
(b) Entropy and 'randomness'.
(c) The Nernst Heat Theorem.
(d) The First Law of Thermodynamics.
(e) The applications of Carnot's Cycle. (C)

58. Mention *three* methods by which the heat of reaction at constant pressure may be measured. What further information is required in order to calculate the free energy change? Indicate how this additional information may be obtained. (C)

See also questions P. 8, 9, 10, 16; Sp. 9; X. 5, 20, 21, 23; and Section E.

Notes. In thermochemistry Hess's law takes the place of the first law of thermodynamics and appears frequently on question papers, often with calculations. The latter are usually quite straightforward provided that the correct signs are allocated to all heat changes. Always use the correct thermodynamic sign convention, namely, that all changes are considered from the point of view of the system, so that heat added to the system (absorbed) is positive and heat lost by the system (evolved) is negative.

Question 1 contains serious errors. The accepted values of the heats of combustion are -138 kcal for ethane, -68 kcal for hydrogen, and -94 kcal for graphite. While the incorrect values given in the question do not affect the method of calculation they do lead to extremely high values of the heats of formation of ethane. A student familiar with heats of formation of organic compounds might be misled into loss of confidence in his calculations.

Question 5 emphasizes that the Born–Haber cycle is one application of Hess's law, and also the importance of the relationship between heats of combustion and of formation and bond energies.

The calculation of heats of reaction at constant pressure from those at constant volume (or vice versa) makes use of the ideal gas equation by substituting ΔnRT for $P\Delta V$. It is therefore only an approximate method since it neglects volume changes in solids and liquids and assumes that all vapours or gases behave ideally. The term Δn is then the change in the number of moles of gases or vapours during the reaction and must carry the appropriate sign.

Heat capacities are extremely important in physical chemistry. The questions here are typical of many on the subject and should well repay careful consideration. A glance through any textbook will reveal several examples of the application of heat capacities in addition to those suggested by the questions themselves. Make a summary of as many applications as you can find.

Answers to questions on heat capacities must not be restricted to those of gases unless the question is so limited, and, in any case, reference must be made to actual (non-ideal) gases in addition to ideal gases. The variations of C_p and C_V and of their difference and ratio for non-ideal gases should be clearly related to the complexity of the gas molecules.

The group of questions on chemical equilibria emphasizes the importance of a thorough knowledge of the criteria of equilibrium. These are based on the second law of thermodynamics through the concept of reversibility, and the fundamental (Gibbs) criteria are expressed in terms of entropy and energy changes in an isolated system.

Note the frequent occurrence of simple calculations involving one of the various forms of the Clapeyron equation or the Clausius–Clapeyron equation. Avoid confusion of units; application of the method of dimensions as a check on the use of the correct units is strongly recommended. Note particularly that when the pressure is expressed in atmospheres the heat term must be in litre-atmospheres or c.c.-atmospheres.

In the calculation of equilibrium constants by means of the van't Hoff isochore one may assume a constant value of the enthalpy change *only* over a small temperature range. Otherwise (question 18) the actual value of the enthalpy change must be calculated for each temperature concerned.

For the calculation of question 19 one has (for each temperature) the relationship between the standard free energy change and the equilibrium constant, and also the integrated form of the van't Hoff isochore. From the latter the mean value of the standard enthalpy may be found, and using this with the mean of the standard free energy changes in the equation

$$\Delta G° = \Delta H° - T\Delta S°$$

readily gives the mean value of the standard entropy change for the reaction. This method also answers the second part of question 21.

The first part of question 24 may appear rather involved, but careful reading will reveal that it refers to the Clapeyron equation, a conclusion which is supported by the nature of the calculation. The fundamental equations of thermodynamics may be taken to be those summarized in the diagram in our list of equations. The next five questions are also concerned with the Clapeyron or the Clausius–Clapeyron equations.

The fallacy that a gas expanding through a porous plug is always cooled is exploded in question 42. The direction of the heat change depends on whether the initial temperature is above or below the 'inversion temperature'.

Do not jump to the conclusion that the partial molar properties of a component will be the same in an ideal solution as in the pure state (question 46). Changes in enthalpy and entropy on forming the solution will prevent this. Even so, a little thought will show for what kind of solution the two partial molar properties will coincide.

W. WAVE MECHANICS AND QUANTUM THEORY

QUANTUM theory and statistical mechanics are conveniently included with wave mechanics here. The short list of definitions does not reflect the relative importance of the subject but is largely due to its mathematical nature.

(i) Definitions

Sigma molecular orbital.
Pi molecular orbital.
g and u types of molecular orbitals.
Hybridization of atomic orbitals.
Partition function.

(ii) Laws and Equations

The Heisenberg uncertainty principle.
De Broglie's equation.
The Schrödinger wave equation.
The Pauli exclusion principle.
Hund's rules.

(iii) Questions

1. Discuss the way in which wave-mechanical ideas and calculations have contributed to our understanding of the chemical bond. (Ge)
2. Discuss the experimental evidence which shows that electrons and photons can behave as either waves or particles. How does wave-mechanics reconcile these apparently conflicting types of behaviour? (Ge)
3. Discuss briefly the evidence for the wave nature of electrons. Show in outline, with suitable examples, how wave-mechanical ideas lead to the result that energy is quantised. (Ge)
4. Give an account of the discoveries which led to the introduction of first the Bohr Quantum Theory and secondly Schrodinger's Wave Mechanics. Explain the existence of discrete energy levels in terms of *either* Bohr's *or* Schrodinger's theory. (Gi)
5. In terms of molecular orbital theory, discuss the factors which contribute to the polarity of chemical bonds in simple molecules. Illustrate your answer by reference to examples. (Gi)
6. *Either* (a) Write down the Schrödinger wave equation, defining the symbols used. Indicate how this equation can be applied to the solution of any one problem of chemical interest:
   Or (b) Discuss the principles underlying the determination of crystal structure by the use of X-ray diffraction. With reference to simple examples show how the experimental results can be interpreted. (R)
7. Write down the Schrödinger equation for the hydrogen atom and show that the wave-function $\psi = b e^{-r/a}$ is a solution if a and b are constants and r is the distance of the electron from the nucleus. Evaluate a and b, and discuss the information which can be obtained from this wave-function. (S)
8. Explain what is meant by the de Broglie wavelength of a particle and show how this concept may be used to derive (a) Heisenberg's uncertainty principle for position and momentum, and (b) Schroedinger's equation (amplitude equation). Describe briefly what information about the behaviour of a particle may be deduced from its wave-function. (S)
9. Explain the significance, for the development of the quantum theory, of work on black-body radiation and the photo-electric effect. (S)
10. Discuss the specific heats of monoatomic and diatomic gases in terms of statistical mechanics. (S)
11. Give a quantum mechanical discussion of the bond in the hydrogen molecule-ion H_2^+. (S)
12. State the conditions which must be satisfied by an acceptable solution of the wave equation. Show, without full mathematical details, how these conditions may be applied to derive the permitted energy

states of a simple harmonic oscillator. Discuss the information which may be deduced from the result. (S)

13. *Either* Describe the distribution of the electrons forming the bonds between the two carbon atoms in C_2H_6, C_2H_4, and C_2H_2, and discuss the properties of these molecules in terms of these distributions.

Or Discuss directed valency. (Gi)

14. Select *either two* of the following for brief discussion, *or one* of the following for more detailed discussion:

(a) directed valence;
(b) non-localised molecular orbitals;
(c) the ionic bond;
(d) the hydrogen bond. (Ge)

15. Discuss the nature of the chemical bond in diatomic molecules. (Ge)

See also questions in Sections Sp and M.

Notes. The basic idea to be developed in answer to question 1 is the 'new' view of the electron. In order to apply the concept to chemical bonds it is necessary to explain atomic orbitals and their linear combination to form molecular orbitals. Either the 'electron charge distribution' or the statistical picture may be used in explaining orbitals, but any method of representation must of necessity be simplified; for example, p-orbitals are *not* 'hour-glass' or 'dumb-bell' shaped. The electron charge falls off exponentially to zero at infinity.

The statement in question 3 that 'energy is quantised' requires qualification. It should also be noted that whereas translational energy of molecules is often considered to be non-quantized a more correct view is that there is quantization but that the energy levels are extremely close together. One might start the answer to this question from the simple consideration of the 'particle in a box', in which a confined particle is treated as a *stationary* wave. By considering the amplitude rather than the displacement of such waves and extending the ideas to three-dimensional wave motions one might demonstrate the effects of the quantization of the energy of orbital electrons. Plotting the function $4\pi r^2 \psi^2$ against r will be useful here.

Polarity of bonds (question 5) arises from an unsymmetrical charge distribution of bonding electrons and is readily explained by means of overlapping orbitals. Examples must include homonuclear and heteronuclear diatomic molecules with overlapping of similar and of dissimilar orbitals.

X. MISCELLANEOUS

In this section we have collected together questions which cannot be included conveniently under the previous topics. They include questions on topics which arise infrequently in papers of Part I standard, and also 'mixed' questions, particularly those in which a wide choice of topics is given. For the latter group the appropriate definitions, laws, and equations will have been given previously.

Although the questions are not sub-divided it will be seen that several on molecular weight determinations appear first, followed by a few on gases, liquefaction and related subjects, and finally the mixed questions.

(i) Definitions

Ideal gas.
Ebullioscopic constant (molecular elevation constant).
Cryoscopic constant (molecular depression constant).
Limiting density.
Critical temperature, pressure, and volume.
The Boyle temperature.
Cohesive energy density.
Cybotactic groups.

(ii) Laws and Equations

Raoult's law.
Henry's law.
Dalton's law of partial pressures.
Van der Waals's equation.
The virial equation.

> These two equations of state are considered to be essential, but some acquaintance with two or three others might be useful. The virial equation is the most commonly used equation of state, and the significance of the first and second virial coefficients should be known.

Graham's law.
The law of rectilinear diameters.
The law of corresponding states.
The expression for the elevation of the boiling point.
The expression for the depression of the freezing point.
The osmotic pressure equation.
The relationship between osmotic pressure and vapour pressure.

The equation for the molecular weight of a gas from its limiting density.
Kinetic theory expressions for:
  The product PV;
  The kinetic energy of a gas;
  The mean free path of gas molecules;
  The viscosity of a gas.
The principle of the equipartition of energy.
Trouton's rule.
The Eötvös equation.
Stokes's law.
Poiseuille's equation for the viscosity of a liquid.
The Staudinger equation.
The law of rational indices.
The Bragg equation.

(iii) Questions

1. Write a critical survey of the methods available for the determination of the molecular weight of a substance in solution. Comment briefly in each case on any particular experimental or theoretical difficulties which might be encountered in obtaining an accurate result. (R)
2. Give an account of the methods which can be used for the determination of molecular weights in solution, emphasising the theoretical principles upon which the methods are based. (Gi)
3. Write a brief account of the methods available for determining molecular weights in solution.
   (The derivation of thermodynamic formulae is not required, but the advantages and limitations of each method should be mentioned, as well as the factors that would influence you in a choice of method and solvent.) (R)
4. Discuss the principles underlying the determination of the molecular weight of a substance by the boiling-point-elevation method. Describe briefly how such measurements may be carried out in the laboratory. (R)
5. State Raoult's law and discuss its application to the determination of molecular weights by the freezing point depression method. (R)
6. Write a short account of *either* cryoscopic *or* ebullioscopic methods of determining the molecular weight of substances in solution. Briefly discuss the relative advantages and disadvantages of the two methods. (R)
7. A is a weak organic acid, soluble in water. B is a stable gas with a critical temperature of $-94°$ C. Describe, for each of these substances, a method of determining its molecular weight accurately (i.e. to better than 0·5 per cent). (R)

X. MISCELLANEOUS

8. Suppose that you are provided with a small, well-formed crystal of a substance, with a request to determine its molecular weight. The crystal is so small that a micro-method must be used. Mention two methods and describe one of them in some detail. (R)

9. Derive a relationship between the osmotic pressure of a solution and the partial pressure of the solvent. Explain clearly the conditions under which a relationship between the osmotic pressure of the solution and the concentration of the solute may thence be deduced. Review briefly the application of osmotic methods to the determination of molecular weights. (S)

10. (a) One litre of air at $15 \cdot 4°$ C is bubbled through a pure sample of benzene (C_6H_6) and becomes saturated with the vapour. The pressure is one atmosphere throughout. Calculate the weight of benzene removed from the liquid on the assumption that the system is an ideal one. The vapour pressure of benzene at $15 \cdot 4°$ is $60 \cdot 0$ mm.

(b) At $26 \cdot 7°$ C a sample of N_2O_4 is partly dissociated into NO_2 molecules, and the measured vapour density ($H = 1$) is $38 \cdot 3$. Calculate the volume percentage of NO_2 in the gas. (R)

11. Explain what is meant by an equation of state and outline the principles upon which the van der Waals equation is based. Show how this equation may be written as a reduced equation of state and discuss the advantages of this form of equation. (R)

12. Give an account of the use of the van der Waals equation to interpret the properties of real gases, emphasising its successes and its failures. Discuss briefly the reasons why the boiling points of argon, hydrogen chloride, water and sodium chloride increase in this order.

(S)

13. The liquefaction of a vapour takes place in response to forces of attraction which act between the molecules of the vapour, but for each substance the process of condensation can be observed to occur only over a finite range of pressures and temperatures. Discuss this statement. (R)

14. Write an essay on Avogadro's number. (Gi)

15. Describe one method each for determining the values of the gas constant, **R**, and the Avogadro number, **N**. (R)

16. Discuss the purification of chemical substances by (a) distillation, (b) crystallisation, and (c) any *one* method which has become available within the last twenty years. (S)

17. Show how developments in the definition of acids and bases and in our understanding of ionic solutions have contributed to the interpretation of acid–base catalysis. (Ge)

18. The presence of an indifferent electrolyte can affect the colour of an indicator, the rates of certain catalysed reactions, or the solubility of a sparingly soluble salt. Write a general account of 'neutral salt effects' such as these, and explain how they arise. (R)

19. Explain what is meant by the terms 'ideal solution' and 'activity'. Discuss the application of the concept of activity to non-ideal solutions.

(R)

20. Write an essay on *one* of the following:

   (*a*) dipole moments;
   (*b*) ionic mobilities;
   (*c*) calorimetry;
   (*d*) the liquefaction of gases. (R)

21. Write on *two* of the following:

   (*a*) the critical state;
   (*b*) the accurate determination of gas densities;
   (*c*) the Principle of Equipartition of Energy;
   (*d*) the Joule–Thomson effect. (R)

22. Write informative notes on *two* of the following:

   (*a*) the critical state;
   (*b*) the specific heats of gases;
   (*c*) osmotic pressure;
   (*d*) the Avogadro number. (R)

23. Write informative notes on *two* of the following:

   (*a*) the van der Waals equation;
   (*b*) the Joule–Thomson effect;
   (*c*) the Kohlrausch law of independent mobilities;
   (*d*) the determination of the charge of the electron. (R)

24. Write briefly on *two* of the following:

   (*a*) Molecular refractivity and chemical constitution.
   (*b*) The variance of polycomponent systems.
   (*c*) The principle of equipartition of energy.
   (*d*) The determination of Avogadro's number. (R)

25. Write notes on *three* of the following:

   (*a*) buffer solutions;
   (*b*) acid–base indicators;
   (*c*) adsorption indicators;
   (*d*) the charges on colloid particles;
   (*e*) the precipitation of colloids by electrolytes;
   (*f*) peptization. (R)

26. *Either* write an essay on *one* of the following topics, *or* give a brief

X. MISCELLANEOUS

account of *two* of the following topics, in either case giving particular reference to analytical applications:

(a) adsorption chromatography;
(b) partition chromatography;
(c) gas–liquid chromatography;
(d) polarography. (Ge)

27. *Either* write an essay about *one or* give brief accounts of *two* of the following, with particular reference to their use in analytical chemistry:

(a) partition chromatography;
(b) zone electrophoresis;
(c) polarography;
(d) absorption spectrometry. (Gi)

28. Write a brief essay on *one* of the following topics:

(a) The stability of colloidal solutions;
(b) Electrokinetic phenomena;
(c) The mechanism and kinetics of the synthesis of some common polymer. (Gi)

29. Write an essay on *one* of the following:

(a) specific heats of solids;
(b) colloidal solutions;
(c) heterogeneous catalysis. (Gi)

30. Write an account of

Either (a) the interpretation of acid–base phenomena in aqueous and non-aqueous solvents in terms of the Brønsted-Lowry concept;
or (b) the principles of methods available for studying the formation, formulae and stability constants of complex ions in solution. (Ge)

31. Write briefly on *two* of the following:

(a) spectroscopic ground states of atoms;
(b) the Principle of Equipartition of Energy;
(c) the surface tensions of liquids;
(d) the Joule–Thomson effect;
(e) X-ray diffraction. (S)

32. Give an account of analytical procedures based on *two* of the following:

(a) adsorption;
(b) light absorption;
(c) ionization and dissociation brought about by electron impact. (Ge)

33. Write short notes on *three* of the following:

(a) The Beer–Lambert Law;
(b) The quantum-efficiency of a photochemical reaction;
(c) The ion-atmosphere (Debye–Hückel Theory);
(d) Ampholytes;
(e) The classification of colloidal systems. (D)

34. Explain *four* of the following:

(a) Why solid carbon dioxide can be kept for some time at N.T.P.
(b) 'Retrograde solubility.'
(c) The kinetics of the conversion of para-hydrogen to ortho-hydrogen.
(d) The use of the term 'component' in the Phase Rule.
(e) 'Incongruent melting point.'
(f) The relationship between K_p and K_c in a gas reaction. (C)

See also questions E (f) 4; S. 27; T. 41, 43; and Section P for questions on Raoult's and Henry's laws.

Notes. It will be noticed that most questions on the determination of molecular weights in solution emphasize the critical approach. Examiners tend to conclude that failure to appreciate the difficulties of technique or advantages and limitations of the various methods indicates that a student has not actually carried out the experiments concerned in the laboratory. This might sometimes be true, but the limitations of a method generally appear from a consideration of the underlying theory, and the practical difficulties can usually be foreseen by anyone with general laboratory experience. What is so often lacking is practice in expressing such matters; in other words, the critical faculty has not been developed throughout the course of instruction.

An important point which must be considered is the choice of solvent. Water is probably one of the least suitable solvents for the determination of molecular weights. One must consider the general convenience, physical properties, and, in particular, the size of the depression or elevation constant of a proposed solvent. The theory gives an expression for the constant from which the factors governing its size may be seen. Hence, the choice of solvent is not entirely a matter of trial and error.

Question 5 requires the derivation of the expression for the molecular weight of a solute from freezing-point depression. Raoult's law is evidently involved, but make sure to use it in conjunction with the *correct* form of the Clausius–Clapeyron equation.

The comment above concerning water as solvent should be kept in mind when considering question 7. Furthermore, a weak organic acid will be partly dissociated in water, so that the ebullioscopic method for

X. MISCELLANEOUS

its molecular weight is ruled out. Mention of the solubility does, nevertheless, contain a clue as to the method to be used.

In question 9 we meet the well-known theme that one may deduce thermodynamically various relationships between vapour pressure and other physical properties, but that in relating these to the concentration of a solution we must make use of the empirical relationship of Raoult's law. In this way we introduce limitations to the conclusions reached.

Andrew's isothermals will provide a suitable diagram to help in the explanation of the phenomena of liquefaction of vapours (question 13).

There are several recognized methods for determining R and N; it is *not* permissible to select at random one of the innumerable equations in which one of these constants occurs, insert the values of all other factors, and hence calculate the value of R or N. It might be a useful exercise to discuss the objections to such a procedure.

To be asked to discuss the purification of substances by distillation and by crystallization in fifteen minutes each may appear a tall order. Only the briefest outline is possible, avoiding any side issues and concentrating on purification. Show what kinds of substances may be purified by each method and the relative advantages of each, and give some idea of the degree of purity attainable. Corresponding facts are required concerning the selected modern method, which might be chromatography, ion exchange, zone-melting, etc.

Questions 18 and 19 are concerned with 'activity' and how the activity of a solute is affected by the presence of other substances. In the former question all the solutions concerned are ionic and the effects may be explained by the use of the ionic strength and the Debye–Hückel limiting equation. Question 19 covers both ionic and non-ionic solutions and may be answered with more emphasis on thermodynamic aspects.

The remaining questions in this Section are 'mixed' questions; in case of difficulty turn to the appropriate Section of Chapter 2, where there will most probably be notes on questions on a similar topic.

3

Annotated Full Answers to Selected Questions

THE questions to be fully answered here have been selected so as to provide examples of various styles of answer (*see* Section 1.3) and to illustrate the important points raised in Chapter 1.

The numbers in parentheses are *not* part of the answers, but refer to the notes to be found at the end of each answer. As mentioned in the Introduction, students are advised to write out their own answers, either in full or in outline, before referring to a full answer in this chapter.

Question E(a).1 (page 18)

(1) (2) The transport number of an ion (t_i) is the fraction of the total current carried by that ion. It is also the ratio of the ionic conductance (or ionic mobility) (λ_i) to the total equivalent conductance (Λ) of the electrolyte, or, in other words, the ratio of the contribution of the ion to the total conductance of the electrolyte solution, i.e.

(3) $$t_i = \lambda_i/\Lambda$$

Also, since $\lambda_i = u_i F$, where u_i is the absolute mobility of the ion (i.e. its speed under one volt per centimetre potential gradient) and F is the Faraday,

$$t_i = u_i/(u_i + u_j)$$

(4) This relationship, which applies only to a solution of a uni-univalent electrolyte, suggests that the transport number might be found from measurements of absolute mobilities. This is in fact so, the methods used being known as 'moving boundary' methods.

The movements of ions under a potential gradient lead to concentration changes in electrolytic cells, and measurement of such changes near the electrodes was used by Hittorf to determine transport numbers. A third method is by the use of certain galvanic cells in which the e.m.f. depends on the transport number of an ion moving through the cell.

(5) From the last equation above it may readily be shown that, for inert electrodes,

$$t_+ = \frac{\text{Number of equivalents lost from anode region}}{\text{Number of Faradays passed}}$$

and $$t_- = \frac{\text{Number of equivalents lost from cathode region}}{\text{Number of Faradays passed}}$$

Hittorf's apparatus consists of a cell of three compartments which may be separately drained for analysis after electrolysis. There must be no change of concentration in the centre section in order to ensure that all concentration changes have been confined to the solutions near the electrodes. The quantity of electricity passed through the cell must be accurately measured. Corrections are necessary, in particular for the transport of water by hydrated ions.

(6) If it were possible to follow the movements of ions through a solution during electrolysis then, from their speeds, the transport numbers could be found. This is the basis of moving-boundary methods. Ideally the boundary to be observed is between coloured solutions in a narrow tube or a gel. The transport number of an ion may be calculated from the volume swept through by the boundary during the passage of a known quantity of electricity. When the solutions are colourless, indicators may be used or changes in refractive index followed.

The main difficulties in moving-boundary methods are associated with forming and retaining sharp boundaries.

(7) When e.m.f. methods are available for a determination they are generally found to be the most accurate and often the most convenient methods. This is so for the determination of transport numbers, for which the method will be described in detail.

(8) If the same reaction can be carried out in a concentration cell with transport and in a concentration cell without transport, giving e.m.f. E_t and E, respectively, then

$$E_t = 2t_-\, RT/F \ln \frac{(a_\pm)_1}{(a_\pm)_2},$$

for cation electrodes and a uni-univalent electrolyte,

and $$E = 2\, RT/F \ln \frac{(a_\pm)_1}{(a_\pm)_2}.$$

Hence, $E_t/E = t_-$.

(9) But a concentration cell is one in which the two electrode systems differ only in the concentration of one species, and transport numbers vary with concentration. This means that the value determined as above will be a mean value. It is therefore necessary to make several measurements with different concentrations in one half-cell and plot E

and E_t against log $(a_\pm)$. Then t at any concentration may be calculated from

$$t = \frac{\mathrm{d}E_t/\mathrm{d}\log a}{\mathrm{d}E/\mathrm{d}\log a}.$$

(10) It is not always easy to devise suitable cells for this method, particularly those without transport. A salt bridge between the two solutions is not satisfactory; it will not entirely eliminate transport. For accurate work the junction itself should be eliminated, for example by combining two similar cells in opposition, as follows:

$$H_2 \text{ (1 atm)}|HCl\ (a)_1, AgCl(s)|Ag|AgCl(s),\ HCl(a)_2|H_2 \text{ (1 atm)}$$

The corresponding cell with transport would have the two HCl solutions in contact within a porous plug or similar device to prevent appreciable diffusion, or a flowing junction might be used.

(11) The e.m.f. of the cells are measured by means of a potentiometer, which may be based on the Poggendorff compensation method or might be a direct-reading electronic millivoltmeter. The basic circuit of the former is shown in the diagram (Figure 11).

The galvanometer G shows zero current when the slide is at positions C_s or C_x on the potentiometer wire with the standard cell or the test cell, respectively, in the circuit. Then the e.m.f.s are related by

$$\frac{E_x}{E_s} = \frac{AC_x}{AC_s}$$

(12) One of the definitions of transport number given above shows its relationship to conductance, namely

$$\Lambda t_i = \lambda_i$$

The equivalent conductance is calculated, for a cell with fixed electrodes, from

$$\Lambda = k\frac{C}{c}$$

where k is a constant determined from measurements on a standard potassium chloride solution, C is the measured conductance in ohm^{-1}, and c is the equivalent concentration. Hence, (13) from measurements of transport numbers and equivalent conductances one may determine individual ionic conductances.

Notes. (1) This is a 'review' question; we choose to answer in essay form, though this is not essential (Section 1.3). The answer given occupied a little over four sides of quarto paper in manuscript, which is a reasonable length for thirty-five minutes' work.

(2) The subject is introduced by giving two definitions (Section 1.3, last paragraph).

ANNOTATED FULL ANSWERS TO SELECTED QUESTIONS 89

(3) Related functions are mentioned, which will be useful to refer to at a later stage.
(4) Note how the methods are introduced as logical consequences of the theoretical relationships already given.
(5) Hittorf's method is briefly explained and a difficulty mentioned.
(6) Moving-boundary methods are similarly treated. More time could be given to this section if available.
(7) Introduction and detailed description of the e.m.f. method, and reasons for the choice of this method. One is not expected to give reasons for a choice of method or question, but when there is a good reason one might hope to gain a mark or two by briefly stating it.

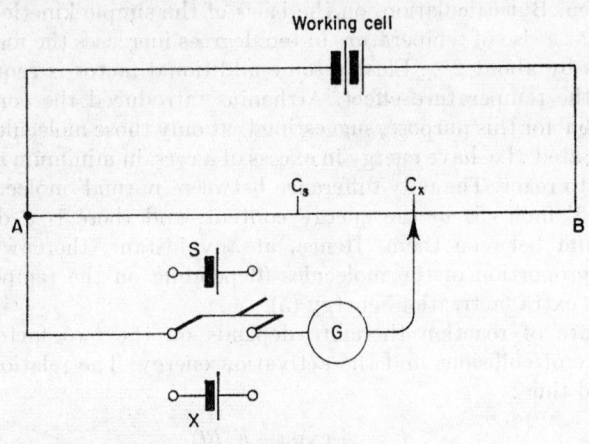

Figure 11

(8) Basic theory of the method.
(9) Development of the method.
(10) Difficulties in the use of the e.m.f. method.
(11) Practical details of the method, including diagram of the electrical circuit. The diagram must be simple and clear (Section 1.6). If time permitted, more details of a modern potentiometer could be given, and brief mention of the basic principles of an electronic instrument.
(12) The answer to the last part of the question also follows directly from one of the equations given earlier.
(13) The answer is rounded off by a short summarizing sentence.

Question K.2 (page 34)

(*See*, first, note (1), page 93)
(2) Chemical reactions are of three kinds: homogeneous, heterogeneous, and chain reactions. There are several factors which influence the rates of all reactions. These will be considered first, followed by some of the

additional factors which affect the rates of heterogeneous and chain reactions.

1. *Temperature* (3)

The rates of all chemical reactions are affected by temperature changes (4). A rise of temperature of ten degrees will generally about double the rate of a homogeneous reaction.

Early theories of reaction kinetics interpreted reactions as the result of collisions between molecules. Rise of temperature will increase the number of such collisions and so would be expected to increase the rate of reaction. But calculation, on the basis of the simple kinetic theory, shows that a rise of temperature of ten degrees increases the number of collisions by about 2%. Clearly some additional factor is required to explain the temperature effect. Arrhenius introduced the concept of 'activation' for this purpose, suggesting that only those molecules which are 'activated', i.e. have energy in excess of a certain minimum amount, are able to react. The only difference between 'normal' molecules and 'activated' moles is in the energy content, and there is a dynamic equilibrium between them. Hence, at any instant, there will be a definite proportion of the molecules (depending on the temperature) with this extra 'activation' energy (5).

The rate of reaction therefore depends on the two factors: the frequency of collisions and the activation energy. The relationship is expressed thus:

$$k = A \exp(-E/Rt)$$

where k is the rate constant,

A is the 'frequency factor',

E is the activation energy,

R is the gas constant per mole,

and T is the absolute temperature,

$$\text{or } \frac{d \ln k}{dT} = \frac{E}{RT^2} \quad (6)$$

According to the collision theory of reactions the frequency factor should be equivalent to the 'collision number', i.e. the number of molecules taking part in collisions in one cubic centimetre in one second. This is sometimes found for bimolecular reactions, although the rate is often less than expected. In such cases a 'steric factor' or 'probability factor', which may be between 1 and 10^{-8}, is also necessary. The reason for this factor is explained by the Activated Complex or Transition State theory.

2. Pressure

The effect of pressure is appreciable only in the case of gas reactions, where pressure may be considered analogous to concentration in reactions in solution.

In the case of unimolecular gas reactions the effect of pressure is very important. Here the number of collisions occurring depends on the pressure of the gas, but the time required for the pressure to be reduced to half its initial value is independent of that initial value. Hence a unimolecular gas reaction is not the *direct* result of collisions.

Lindemann explained the effect by postulating a time lag between activation and decomposition of a molecule. There are then two processes: the activation equilibrium and the decomposition reaction. By combining these he showed that the rate of reaction is directly proportional to the pressure of the gas, i.e. the reaction is of the first order. But, at low pressures fewer molecules become activated by collisions and the delay between activation and decomposition is equal to that between activation and de-activation by collisions. Below this pressure every molecule becoming activated will decompose, and the rate becomes proportional to the square of the pressure, i.e. the reaction becomes of the second order (7).

3. Concentration (8)

The effect of concentration on the rate of a reaction is expressed in terms of the order of the reaction, defined as 'the number of species whose concentrations determine the rate of the reaction'.

For a first-order reaction, such as the thermal decomposition of cyclobutane, the rate is expressed in terms of the 'integrated rate equation'

$$k = 1/t \ln(a/(a-x)),$$

relating the rate constant (k) to the initial concentration (a) and the reduction in concentration (x) after a time t.

More complex equations may be deduced for second- and third-order reactions. An interesting and useful consequence of the order of reaction and its relation to concentration is the idea of the 'time of half-change'. In a first-order reaction the initial concentration is reduced to half its value in a time equivalent to $\ln 2/k$, i.e. independent of the initial concentration. For second- and third-order reactions in which the initial concentrations of all reactants are equal to a, the times of half-change are $1/ka$ and $3/2ka^2$ respectively.

4. Other Factors

Reaction rates may also be affected by catalysts or by the absorption of radiation. The latter is studied as photochemistry (including radiation chemistry), but time does not permit discussion of these topics here (9).

5. *Heterogeneous Reactions* (10)

The commonest kind of heterogeneous reaction, that of a gas adsorbed on a solid surface, is, in effect, a catalysed reaction. The first successful treatment of the kinetics of such reactions was by Langmuir in 1916. For a single reacting gas he deduced a rate equation of the form

$$\frac{dx}{dt} = \frac{k_1 p}{1 + k_2 p}$$

where p is the pressure of the gas and k_1 and k_2 are constants.

This relates the rate of reaction to the pressure of the gas, and the theory is readily extended to cover two or three reacting gases and the effects of strong or weak adsorption of reactants or products on the solid surface. The above equation shows that, even in the simplest case, the reaction is not of the first order unless $k_2 p \ll 1$.

The temperature effect in such reactions may be explained by means of the Arrhenius equation, but the activation energy may be increased or reduced by one or more heats of adsorption.

6. *Chain Reactions*

These form a very important class of very fast consecutive reactions, which may also involve alternative routes or reactions ('branched chains'). A single chain of reactions will include an initiation reaction, propagation reactions, which may be repeated many times, and finally a chain-breaking reaction. These may be represented as follows for the reaction between hydrogen and chlorine:

$Cl_2 = Cl + Cl$ Initiation reaction

$\left. \begin{array}{l} Cl + H_2 = HCl + H \\ H + Cl_2 = HCl + Cl \end{array} \right\}$ Chain propagation

$\left. \begin{array}{l} Cl + Cl = Cl_2 \\ H + H = H_2 \\ H + Cl = HCl \end{array} \right\}$ Chain breaking

The effect of concentration on a chain reaction is abnormal, since the concentration determines not only the number of chains started but also how successfully they are propagated.

Other factors which affect the rate of a chain reaction are:

    (*a*) The surface area of the reaction vessel. An increase of surface area slows down the reaction because chain-breaking usually occurs at surfaces.

ANNOTATED FULL ANSWERS TO SELECTED QUESTIONS

(b) Addition of an inert gas speeds up a chain reaction by reducing chain-breaking at the walls.

(c) Inhibitors. The rate may be considerably reduced by the addition of an 'odd-electron' compound such as nitric oxide. These tend to combine with free radicals, which are often involved in chain reactions.

Hinshelwood expressed the rate of a chain reaction by

$$\frac{F(c)}{f(s) + f(c) + A(1 - a)}$$

where $F(c)$ is a function of the concentration of the reactants,
$f(s)$ is a factor for chain-breaking at surfaces,
$f(c)$ is a factor for chain-breaking in the gas phase by collisions,
A is a constant, and
a is the number of species of chain-carrier formed by each chemical act.

(11) It will be seen that a considerable amount of useful information concerning a reaction may be deduced from the effects of these various factors on the velocity of the reaction.

Notes. (1) We have adopted an informal style for this discussion, with the use of sub-headings.

(2) Introduction, indicating how the subject is to be covered. This is a useful approach, since it shows the examiner that the candidate understands what is required, and sets out the writer's aims, even though these might not be fully realized owing to lack of time.

(3) Where practicable the general pattern will be to state the effects of the factor concerned, and then to indicate how far these are explained by appropriate theories.

(4) The only reactions not influenced by temperature are nuclear reactions. As these are not true *chemical* reactions the statement is correct.

(5) The distribution of energies could be discussed here.

(6) The derivation of the Arrhenius equation is relevant but not essential.

(7) The explanation of the effect of pressure on unimolecular gas reactions was a great victory for Lindemann's theory, and provides a popular examination question (e.g. questions K. 15 and 24).

(8) Concentration is probably, in many cases, the most important factor affecting the rates of reactions. This section could well be expanded if time were available.

(9) Again we show that certain factors have not been overlooked.

(10) Concepts developed earlier in a general way are now applied specifically to heterogeneous reactions and then to chain reactions.

(11) Do not leave the subject 'high and dry'. The value of this summarizing conclusion is out of all proportion to its length.

Question T.1 (page 66)

(*See*, first, note (1), page 95.)
(2) Hess's law of constant heat summation may be stated thus:
'The heat absorbed or evolved in a chemical reaction at constant pressure or at constant volume is independent of the path taken, whether in one or more stages' (3).

The following example (4) will illustrate the law:

Carbon and hydrogen may be oxidized to form carbon dioxide and water, thus—

$$C + 2H_2 + 2O_2 = CO_2 + 2H_2O$$

in which the heat evolved may be ΔH_1 calories at a temperature T and constant pressure P. Alternatively, carbon and hydrogen may combine to form methane—

$$C + 2H_2 = CH_4$$

with ΔH_2 calories evolved at the same temperature and pressure. The methane may then be burnt in oxygen—

$$CH_4 + 2O_2 = CO_2 + 2H_2O$$

with evolution of ΔH_3 calories. We then have two different ways of going from state 1 (i.e. $C + 2H_2$) to state 2 ($CO_2 + 2H_2O$), and, according to Hess's law, the heat evolved is the same in each case, i.e.

$$\Delta H_1 = \Delta H_2 + \Delta H_3$$

The importance of this is that one may calculate the heat of a reaction from the values for related reactions. The correct signs must be allocated to the heat changes; the convention is that heat evolved is negative and heat absorbed is positive. In the example above all the heat changes are negative.

(5) The theoretical basis for Hess's law is that the heat evolved or absorbed in a process at constant pressure is the difference between the enthalpies of the system in the two states. Enthalpy (or heat content) is a thermodynamic function whose value depends only on the state of the system. Hence, $\Delta H = H_2 - H_1$, where H_1 and H_2 are the enthalpies of the system in the initial and final states, respectively.

For reactions at constant volume enthalpy is replaced by the function U, the 'internal energy', i.e.

$$\Delta U = U_2 - U_1$$

and $\Delta H = \Delta U + P\Delta V$, where P is the pressure and ΔV is the volume increase.

ANNOTATED FULL ANSWERS TO SELECTED QUESTIONS

The last term in this equation is significant only in a reaction involving a gas or gases where there is a change in the total number of moles of gas. Then $P\Delta V$ is replaced by ΔnRT, assuming that the gases behave ideally, where Δn is the increase in the number of moles of gas during the reaction, and R is the gas constant per mole (6).

Hess's law is one aspect of the first law of thermodynamics, which it preceded, and which may be stated thus:
(7) 'The total amount of energy of an isolated system remains constant. It may change from one form to another (Clausius)'.

Thus, Hess's law relates only to heat changes, whereas the first law of thermodynamics covers *all* forms of energy, being equivalent to the law of the conservation of energy.

Calculation

(8) The thermochemical equations representing the reactions given are:

(a) $C_2H_6 + 3\tfrac{1}{2}O_2 = 2CO_2 + 3H_2O : \Delta H_1 = -373$ kcal.
(b) $H_2 + \tfrac{1}{2}O_2 = H_2O : \Delta H_2 = -94$ kcal.
(c) C (graphite) $+ O_2 = CO_2 : \Delta H_3 = -136$ kcal.

(9) For the formation of ethane from graphite and molecular hydrogen we have:

(d) $3H_2 + 2C$ (graphite) $= C_2H_6 : \Delta H_4 = x$ kcal.

An alternative path to equation (d) is clearly via equations (b) and (c) and the reverse of equation (a). For the appropriate quantities, and using the correct signs, we get

$$\Delta H_4 = 3\Delta H_2 + 2\Delta H_3 - \Delta H_1$$
$$= 3(-94) + 2(-136) - (-373) \text{ kcal}$$
$$= \underline{-181 \text{ kcal}}.$$

(10) The heat of formation of ethane, at constant pressure, from graphite and molecular hydrogen is thus -181 kcal per mole.

As we have already seen,

$\Delta U = \Delta H - P\Delta V$
so that $\Delta U = -181{,}000 + 2RT$ (since Δn is -2)
$= -181{,}000 + 2 \times 1{\cdot}987 \times 298$
$= \underline{-179{,}816 \text{ cal}}.$

Hence, the heat of formation at constant volume is $-179{\cdot}8$ kcal.

Notes. (1) There are five parts to this question. A possible marking scheme might be:

Statement of Hess's law	2 marks
Explanation of the law	4 marks
Relationship to first law	3 marks
Calculation of ΔH	8 marks
Calculation of ΔU	3 marks
Total	20 marks

Thus, nine marks might be available without the calculation. This is merely to show the relative importance of the calculation; *all* parts of any question must be attempted.

(2) The law must be expressed as part of a complete sentence, not as a 'bald' statement.
(3) Refer to Chapter 1. Does this definition satisfy the criteria?
(4) A simple example explains the application of the law.
(5) The full explanation of Hess's law is based on thermodynamics, and this leads naturally on to the next part of the question.
(6) It is convenient to compare the heats of reaction at constant volume and at constant pressure here and to refer back to this when dealing with the last part of the calculation.
(7) In order to discuss the relationship between Hess's law and the first law of thermodynamics it is essential to state the latter. There are many ways of doing this; we have chosen Clausius's statement as being one of the simplest and most appropriate here.
(8) The calculation is designed to discover whether Hess's law is really understood. Calculations must be set out clearly and concisely. First put down in convenient form the data provided.
(9) A convenient statement of the result required.
(10) The result of the calculation is clearly expressed. As mentioned in the notes to Section 2.T(iii), the *heats of combustion used in this calculation are incorrect*. Substitution of the accepted values leads to more reasonable values for the heat of formation of ethane, namely, -24 kcal and $-22 \cdot 8$ kcal, at constant pressure and volume, respectively.

Question X.25 (page 82)

(*See*, first, note (1), page 98)

(a) Buffer Solutions

(2) A buffer solution shows 'buffer action', defined as 'resistance to change of pH' (3). Its effectiveness is measured by the 'buffer capacity', $d(a)/d(pH)$ (or $d(b)/d(pH)$), the slope of the curve of concentration of added acid (or base) against pH.

The diagram (Figure 12) (4) shows various neutralization curves which illustrate the effects of buffer solutions.

Curves 1 and 4 are for the neutralization of strong acid and strong base, respectively. Here the slope is small, i.e. the reciprocal slope, the buffer capacity, is large, over a wide range of added acid or base.

Curves 2 and 5 relate to weak acid and base, and curves 3 and 6 to very weak acid and base, where the salt formed is appreciably hydrolysed. Each partly neutralized acid or base shows buffer action near the middle of the curve where the slope is least.

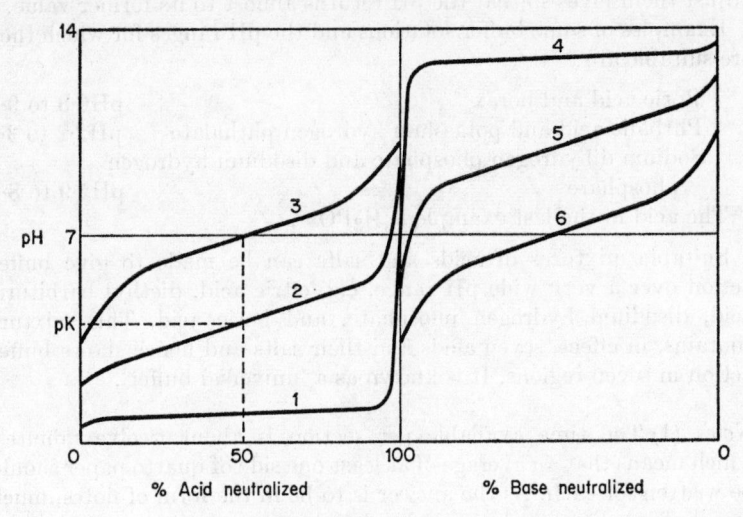

Figure 12

When a weak acid (or base) is 50% neutralized, the pH of the solution is equal to the pK_a of the acid (or $p(OH) = pK_b$ of the base). At this point the solution contains equivalent amounts of acid (or base) and its salt. For other solutions the pH depends on the ratio of the activities of the two, and, in general, for a solution of an acid and one of its salts

$$[H^+] = K_a \frac{[acid] - [H^+]}{[salt] + [H^+]}$$

or, approximately,

$$[H^+] = K_a \frac{[acid]}{[salt]}, \text{ for pH4 to pH10}$$

Or $\qquad pH = pK_a - \log[acid]/[salt]$

If the acid or base is strong, then buffer action is limited to low pH or high pH, respectively. Weak acids or bases are useful for preparing buffer solutions covering a wide range of pH values.

Buffer action of partly neutralized weak acid or base solutions is due to the interaction of the dissociation equilibria of the acid or base and that of water, e.g.

$$CH_3COOH = H^+ + CH_3COO^-$$
$$H_2O = H^+ + OH^-$$

and the salt, say sodium acetate, is completely dissociated. Any addition of hydrogen or hydroxyl ions disturbs these equilibria, which adjust themselves so that the pH returns almost to its former value.

Examples of some buffer solutions and the pH ranges for which they are suitable are:

Boric acid and borax	pH6·8 to 9·2
Phthalic acid and potassium hydrogen phthalate	pH2·2 to 3·8
Sodium dihydrogen phosphate and disodium hydrogen phosphate	pH5·9 to 8·0

The acid in the last example is $H_2PO_4^-$.

Suitable mixtures of acids and salts can be made to give buffer action over a very wide pH range, e.g. citric acid, diethyl barbituric acid, disodium hydrogen phosphate, and boric acid. This mixture contains, in effect, seven acids and their salts and hence shows buffer action in seven regions. It is known as a 'universal buffer'.

Notes. (1) The time available per section is about twelve minutes, which means that an average of at least one side of quarto paper should be written for each. As the answer is to be in the form of notes, much more information can be accommodated on one sheet than in the case of an essay.
(2) Introduction by means of short definitions.
(3) A buffer solution does *not* 'maintain constant pH'.
(4) This diagram is a good example of the type which saves much descriptive work, illustrates several points, and, incidentally, proves the statement in Note (3).

(b) *Acid–base (or Neutralization) Indicators*

(*See*, first, note (1), above.)
(2) *Definition:* A substance whose solution varies in colour according to the pH.

These indicators are always *weak* acids or bases, but they must not themselves appreciably affect the pH of the solution.
(3) *Mechanism:* It was originally assumed that the colour change was due solely to ionization, e.g.

$$HInd. = H^+ + Ind.^-$$
$$\text{Colour 1} \qquad \text{Colour 2}$$

But there is no known reason why this should be so.

It is now believed that the colour change occurs as a result of tautomerism, such as a benzenoid–quinonoid tautomerism. The full reaction goes in stages, thus:

(a) $\quad\quad\quad\quad$ HInd. = H$^+$ + Ind.$^-$
$\quad\quad\quad\quad\quad\quad$ Colour 1 $\quad\quad$ Colour 1

$\quad\quad\quad\quad\quad\quad\quad$ Taut.
(b) $\quad\quad\quad\quad$ Ind.$^-$ = Ind.$^-$
$\quad\quad\quad\quad\quad\quad$ Colour 1 Colour 2

(c) $\quad\quad\quad\quad$ Ind.$^+$ + H$^+$ = HInd.
$\quad\quad\quad\quad\quad\quad$ Colour 2 $\quad\quad$ Colour 2

Combining equations (a) and (b), we have

$$\text{HInd.} = \text{Ind.}^- + \text{H}^+$$
$\quad\quad\quad\quad\quad\quad$ Colour 1 Colour 2

The equilibrium constant for this combined equilibrium is $K_{\text{Ind.}}$, the indicator constant.

(4) *Criteria* of a good indicator are:

(a) It must be a *weak* acid or base.

(b) It must be satisfactory in low concentration, since otherwise it would affect the pH of the solution.

(c) The tautomeric change must be instantaneous.

(d) It must change colour *only* by the tautomeric reaction.

(e) Its colour change must depend *only* on the pH of its surroundings, and not on other factors such as the presence of neutral salts.

No indicator completely satisfies these conditions, but several come very close to doing so.

Quantitative Theory: From the expression for $K_{\text{Ind.}}$ it is easily shown that

$$[\text{H}^+] = \frac{K_{\text{Ind.}} \cdot [\text{HInd.}]}{[\text{Ind.}^-]}$$

Hence, when [HInd.] = [Ind.$^-$], i.e. the indicator has its intermediate colour, then [H$^+$] = $K_{\text{Ind.}}$.

When the indicator is 9% ionic and 91% undissociated acid,

$$[\text{H}^+] = K_{\text{Ind.}} \cdot 91/9 = 10 K_{\text{Ind.}}$$
and $\quad\quad\quad\quad$ pH = $pK_{\text{Ind.}} - 1$

Similarly, when these concentrations are reversed,

$$\text{pH} = pK_{\text{Ind.}} + 1$$

This shows that a change of ±1 pH unit gives a practically complete colour change.

Notes. (1) The alternative name suggests the use of these indicators in neutralization titrations.
(2) The abbreviated definition is acceptable in 'notes'.
(3) A brief outline of the theory of indicators is essential. One might have doubts as to whether to include this or some notes on the choice of indicators for titrations, since time does not permit the inclusion of both. The selection of indicators is a special application, whereas the theory of indicators is fundamental.
(4) There are very many weak organic acids and bases which change colour in acid or alkaline solutions but which would never come near to satisfying these criteria.

(c) *Adsorption Indicators*

Definition (1): A coloured substance used to indicate the end-point of a precipitation titration by its adsorption on the precipitate.

Example (2): the use of fluorescein in the titration of chloride by silver nitrate.

The Electrical Double-Layer (3): formed at the interface between a solid and an electrolyte solution is important here. Either by its own ionization or by the adsorption of ions the solid becomes charged. The firmly bound ions hold oppositely charged ions by electrostatic forces. One layer of these is firmly held near the first adsorbed ions, but the remainder are loosely held in a more or less thick 'diffuse layer'.

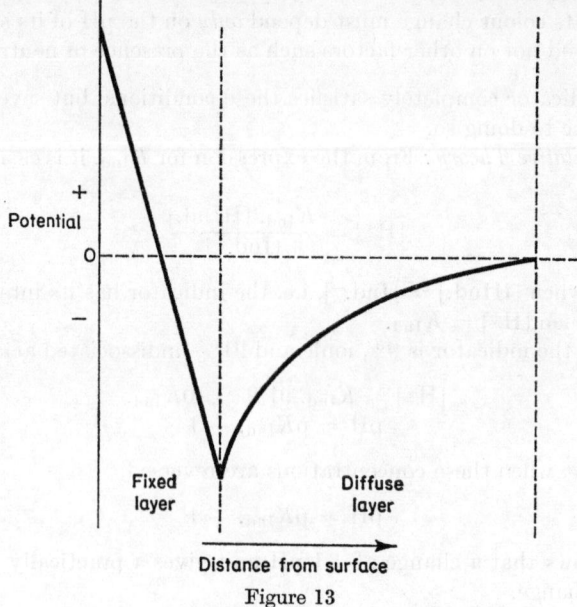

Figure 13

The diagram (Figure 13) (4) shows the resulting potentials. The curve in the diffuse layer might be regarded as showing the decreasing number of excess negative ions (5). In the bulk of the solution there is electroneutrality. The outer adsorbed ions are called 'counter ions'.

Preferential adsorption often occurs. A solid tends to adsorb a common ion, or, from another point of view, to extend its crystal lattice. If there is no common ion available the solid tends to adsorb the one nearest in size.

In the titration of chloride by silver ions the precipitate of silver chloride formed before the end-point, in the presence of excess chloride ions, adsorbs chloride ions. The particles may be represented thus:

$$[AgCl]Cl^-\vdots Na^+$$

the sodium ions being the counter ions. After the end-point there is an excess of silver ions and the adsorption changes to

$$[AgCl]Ag^+\vdots \text{indicator anions}^-$$

The coloured indicator anions (e.g. from fluorescein) are adsorbed as counter ions, so colouring the precipitate.

Notes. (1) Again a definition to begin the answer.
(2) An example is given here so that it may be used to illustrate the mechanism of the process later.
(3) Headings may be used and incorporated in the notes.
(4) The familiar diagram showing swarms of positive and negative ions is difficult to draw clearly and quickly. The potential diagram is much more satisfactory.
(5) This aspect of the potential diagram is worth noting.

4

Sign and Other Conventions

E. ELECTROCHEMISTRY

Cell Diagrams

A glance at the questions in Section 2.E(c) will show that examiners are not always consistent in their methods of drawing cell diagrams. However, the differences are in minor details; there are a few rules which are always observed and which the student should remember, namely:

(a) always indicate the physical state of substances (except pure metal electrodes), and the pressure of a gas, e.g.

$$Hg_2Cl_2(s), \ HCl(aq), \ or \ H_2(1 \ atm)$$

(b) represent the interface between two phases by a single vertical line; but

(c) use a double vertical line to show the effective elimination of the junction potential between two solutions.

Alternative methods are acceptable for certain system. There are, for example, three convenient ways of indicating a second-order electrode (i.e. one for which the electrode reaction consists of two stages). For example, the standard silver/silver chloride electrode may be written:

$$Ag|AgCl(s), \ Cl^-(a = 1)$$
$$Ag, \ AgCl(s)|Cl^-(a = 1)$$
or $$Ag|AgCl(s)|Cl^-(a = 1)$$

When an inert metal is used merely for making electrical contact with the solution, i.e. the metal does not enter into the electrode reaction, there is need of a conventional mode of expression. The questions in Section 2.E(c) reveal the following alternatives:

$$\tfrac{1}{2}H_2(1 \ atm)(Pt)|etc.$$
$$Pt, \ H_2(1 \ atm)|etc.$$
$$H_2(1 \ atm) \ Pt|etc.$$
$$H_2,Pt|etc.$$
$$Pt|Fe^{3+}, \ Fe^{2+}||etc.$$

The last of these is the conventional method for the special case of 'redox' systems, i.e. where both the oxidized and the reduced states of the electrode system are present *together* in the solution. The fourth example is in error in not giving the pressure of hydrogen. The second example is the method preferred by the author, simply because it is convenient to have the substances forming the poles of the cell written at either end of the diagram.

In passing, it should be mentioned that platinum is not always suitable for use as the pole of a gas electrode. Chlorine, for example, attacks platinum, so that it is necessary to use some other material, such as carbon. The standard chlorine electrode is then

$$C, Cl_2(1\ atm)|Cl^-(a=1)$$

Sign Convention for e.m.f.

Most examining bodies in this country have accepted the recommendations of the International Union for Pure and Applied Chemistry agreed at the Stockholm meeting in 1953. The 'IUPAC' or 'Stockholm' convention may be expressed thus:

A cell will be said to have a positive e.m.f. if oxidation occurs at the left-hand electrode and reduction at the right-hand electrode when the cell reaction is proceeding spontaneously.

This has a thermodynamic basis in that when a reaction proceeds spontaneously the free energy change is negative. Also,

$$-\Delta G = zEF$$

so that, for a spontaneous change the e.m.f. of the cell, E, is positive.

It is not essential to refer to the left-hand and right-hand electrodes. The important point is that the sign of the e.m.f. shows the direction of the spontaneous cell reaction. In practice, one *assumes* a direction of reaction, or, what amounts to the same thing, one assumes that oxidation will occur at a certain electrode, and then one calculates the e.m.f. of the cell. If the result is positive, then the assumptions are correct; if negative, then the assumptions must be reversed.

Sign Convention for Electrode Potentials

This follows directly and logically from the sign convention for e.m.f. since single electrode potentials cannot be measured. That is, we can only *compare* electrode potentials, and our standard for comparison is the standard hydrogen electrode.

Theoretically, though not in practice, the potential of an electrode is the e.m.f. of the cell formed by combining the electrode concerned with a standard hydrogen electrode, e.g. the e.m.f. of the cell

$$Pt,\ H_2(1\ atm)|H^+(a=1),\ Cu^{++}(a=1)|Cu$$

is the standard reduction potential of the Cu^{++}/Cu system. As the e.m.f. is found to be positive we conclude that the standard reduction potential of the system is positive. Since the potential of the standard hydrogen electrode is arbitrarily taken to be zero, the e.m.f. is positive if the other electrode is *more positive* than the hydrogen electrode, i.e. if the other electrode is connected to the positive pole of the potentiometer.

Electrode potentials can be expressed either as 'oxidation potentials' or as 'reduction potentials'. One is converted to the other merely by changing the sign. In order to avoid confusion it is recommended that the nature of the potential should be indicated with the symbol, for example, the standard (oxidation) potential of the copper/cupric ion system is $E°_{Cu/Cu^{++}}$, and its standard (reduction) potential is $E°_{Cu^{++}/Cu}$. The subscripts show the direction of the spontaneous electrode reaction.

T. THERMODYNAMICS

The basic convention of thermodynamics is that one must view any changes in a system from the point of view of the system. Hence, anything acquired by the system is positive, and anything lost from the system is negative. This clearly means that heat absorbed by the system is positive and heat evolved is negative.

Students often experience difficulty in understanding the sign of the change when work is done on or by a system. When a system 'does work' on its surroundings it clearly gives up energy to its surroundings, since work is merely the transfer of energy; the transfer is then negative. When work is done on the system by its surroundings the change is positive, since energy is being transferred to the system.

Appendix

RECOMMENDED SYMBOLS

E. Electrochemistry

C measured conductance ($=1/R$).
κ (Greek *kappa*) specific conductance.
ρ (Greek *rho*) specific resistance.
Λ (Greek capital *lambda*) equivalent conductance.
μ (Greek *mu*) molar conductance.
x or α (Greek *alpha*) degree of dissociation.
$\Lambda_\circ$ or Λ_∞ (Greek capital *lambda*) limiting equivalent conductance; equivalent conductance at 'zero concentration' or 'infinite dilution'.
λ (Greek small *lambda*) ionic conductance or ionic mobility.
λ° limiting ionic conductance.
t_+, t_- transport number of cation and anion, respectively.
F the Faraday, which may be taken to be 96,500 coulombs of electricity.
u_+, u_- absolute mobility of cation and anion, respectively.
K an equilibrium constant, its precise nature often indicated by means of a subscript, such as:

K_a acid dissociation constant;
K_b base dissociation constant;
K_W the ionic product for water;
K_H hydrolysis constant.

D (or ϵ) (Greek *epsilon*) dielectric constant.
E an electrode potential or e.m.f., preferably distinguished by the use of subscripts, for example:

$E_{Cu^{++}/Cu}$ the reduction potential of the copper/cupric ion system;
$E_{Cu/Cu^{++}}$ the oxidation potential of the cupric ion/copper system;
E_{cell} the e.m.f. of a cell.

E° a standard electrode potential or standard e.m.f., distinguished by subscripts as above.
n or z the number of equivalents of reaction or number of electrons involved in an electrode reaction.
a activity of a species.
$a_\pm$ mean ionic activity of an electrolyte.
f activity coefficient of a species.

Three activity coefficients are distinguished, namely, f_x, f_c, and f_m (or γ), relating respectively to concentration as mole fraction, molarity, and molality. In the very dilute solutions used in electrochemistry these activity coefficients will be almost identical.

$f_\pm$ mean ionic activity coefficient of an electrolyte.
ν (Greek *nu*) the number of ions produced on dissociation of an electrolyte, being ν_+ cations and ν_- anions.
I or μ (Greek *mu*) the ionic strength of an electrolyte solution.

K. Kinetics

l the mean free path of a molecule.
σ (Greek small *sigma*) the collision diameter.
Z the collision number.
a initial concentration.
x decrease in concentration in time t.
k rate constant, velocity coefficient, or specific reaction rate.
$t_{\frac{1}{2}}$ time of half-change.
A the frequency factor.
E_a activation energy.
P the probability factor or steric factor.

M. Molecular and Crystal Structure

γ (Greek small *gamma*) surface tension
V_m molar volume.
$[P]$ the parachor.
n refractive index.
$[R]$ molar refraction.
μ dipole moment of a molecule.
D or ϵ dielectric constant.
P polarization,

> This may include contributions from three sources, namely:
> P_a atomic polarization;
> P_e electronic polarization;
> P_o orientation polarization.

α polarizability, which, also, may involve α_a, α_e, and α_o.
α optical rotation.

> There should be no risk of confusion between the various uses of this symbol, as they are not likely to be used in the same context.

α_{mag} magneto-optical rotation.
$[\alpha]$ specific rotation.
χ magnetic susceptibility.

P. Phase Rule

P total pressure on a system.
p partial pressure of a gas.
P the number of phases in equilibrium.
C the number of components in a system in equilibrium.
F the number of degrees of freedom (or variance) of a system at equilibrium.

S. Surface Chemistry

Γ (Greek capital *gamma*) surface excess.
α accommodation coefficient.
q differential heat of adsorption.
Q integral heat of adsorption.
W_a work of adhesion.
W_c work of cohesion.
S spreading coefficient.
π film pressure.
V surface potential.
ζ (Greek *zeta*) zeta- (or electrokinetic-) potential.
$[\eta]$ (Greek *eta*) intrinsic viscosity.
D diffusion coefficient.

Sp. Spectra

λ (Greek *lambda*) wavelength.
ν (Greek *nu*) frequency.
$\bar{\nu}$ wave-number ($=1/\lambda$).
R the Rydberg constant.
The quantum numbers—
   for an electron in an atom:

   n principal;
   l azimuthal (or orbital);
   m magnetic;
   s spin;
   j inner.

   for atoms with more than one valency electron:

   L resultant orbital quantum number;
   S resultant spin quantum number;
   J resultant inner quantum number.

for molecules:

J rotational quantum number;
v vibrational quantum number.

Spectral series (atomic spectra):

S sharp;
P principal;
D diffuse;
F fundamental.

> These names are now quite meaningless but of historic interest. Take care not to confuse the various uses of S and s.

T. Thermodynamics

The symbols in brackets are alternatives, those for the thermodynamic state functions being used particularly in the U.S.A. Most examining bodies in Britain now use the symbols first listed.

U (E) internal energy.
F (A) Helmholtz free energy.
G (F) Gibbs free energy.
H enthalpy.
S entropy.
q heat absorbed or evolved.
w work done.
C_p molar heat capacity at constant pressure.
C_v molar heat capacity at constant volume.
f fugacity.
γ (f) activity coefficients (*see* note under f in Section E of this Appendix).
a activity.
K_p equilibrium constant in terms of pressures of gases.
K_c equilibrium constant in terms of concentrations.
K true, or thermodynamic, equilibrium constant.
x or α degree of dissociation.

W. Wave Mechanics and Quantum Theory

ψ (Greek *psi*) wave function.
∇^2 ('del squared') the Laplacian operator.
E total energy of a particle.
V potential energy of a particle.

Index

Note: The first figures in each entry refer to the page number; where a question is concerned the Section letter and the question number follow. For example, 25 E(f).8 means page 25, Section E(f), question 8.

Acid, 29 E(e).1–7; 83 X.30
Activated complex: *see* Transition state
Activation energy, 34 K.4; 35 K.9–11; 36 K.13, 16, 18, 19; 38 K.32; 40 K.39
Activity, 23 E(c).9; 82 X.19
Activity coefficient, 23 E(c). 4, 6; 24 E(c). 16, 17; 25 E(c).18–20
Adsorption, 29 E(e).3; 56 S.1–8; 57 S.19, 23; 83 X.32
Ampholyte, 84 X.33
Anharmonicity constant, 61 Sp.9
Arrhenius theory, 37 K.29
Avogadro's number, 81 X.14, 15; 82 X.22, 24
Azeotrope, 50 P.23; 51 P.29

Base, 29 E(e).1–7
Beer–Lambert law, 60 Sp.1; 62 Sp.17; 84 X.33
Bond energy, 67 T.5
Bond length, 43 M.2, 3; 62 Sp.14
Bond strength, 43 M.3
Brönsted–Lowry theory, 29 E(e).6; 83 X.30
Buffers, 28 E(d).4; 30 E(f).1, 2, 4, 5; 82 X.5

Carnot's cycle, 74 T.57
Catalysis, 36 K.13; 37 K.24; 58 S.27; 83 X.29; 81 X.17
Cells, 22 E(c).1–21; 24 E(c).17
Chain reactions, 34 K.4; 35 K.7, 12; 37 K.26; 38 K.31; 39 K.35; 40 K.39
Chemical potential, 49 P.16; 68 T.13; 69 T.19, 20
Chromatography, 57 S.24; 83 X.26, 27
Clausius–Clapeyron equation, 69 T.24; 73 T.53

Collision number, 38 K.33
Collision theory, 36 K.18
Colloidal electrolytes, 20 E(b).8; 57 S.17, 18, 20
Colloidal state, 56 S.12–15; 57 S.18, 19; 58 S.27, 29; 82 X.25; 83 X.28, 29; 84 X.33
Complex ions, 29 E(e).6; 83 X.30
Conductance, 15; 18 E(a).1, 4, 5; 19 E(b).1; 20 E(b).2–9; 21 E(b).10
Conductometric analysis, 20 E(b).5; 30 E(f).2
Critical state, 71 T.39; 82 X.21, 22

Debye–Hückel theory, 20 E(b).5; 25 E(c).19
Definitions, 6
Diagrams, 9
Diffusion, gaseous, 38 K.33
Dipole moment, 44 M.8–11; 61 Sp.11; 82 X.20
Discussion, 6
Dissociation constant, 20 E(b).8, 9; 25 E(c).20; 28 E(d).3; 29 E(e).1, 5; 31 E(f).7, 9
Distillation, 47 P.2, 4
Distribution, 48 P.9
Donnan equilibrium, 57 S.20, 21
Double layer, electrical, 56 S.11; 57 S.21; 58 S.29

Effusion, 38 K.33
Einstein's law, 62 Sp.17
Electrodes, 22 E(c).1; 31 E(f).8
Electrokinetic effects, 56 S.11; 57 S.25; 83 X.28
Electron affinity, 44 M.11
Electronic charge, 30 E(f).3; 82 X.23
Electrophoresis, 56 S.16; 57 S.20; 83 X.27

INDEX

Electrophoresis effect, 20 E(b).5
E.m.f., 22 E(c).2, 3; 23 E(c).4–6, 8; 24 E(c).10, 15–17; 25 E(c).18–21
Enthalpy, 24 E(c).12; 71 T.30, 34
Entropy, 24 E(c).12; 69 T.21; 70 T.25; 71 T.32, 35, 36; 72 T.40; 73 T.48, 54; 74 T.57
Equation of state, 81 X.11, 12
Equilibrium, 68 T.12–15; 69 T.22
Equilibrium constant, 25 E(c).20, 21; 68 T.12; 69 T.17–20; 73 T.49; 84 X.34
Equipartition of energy, 68 T.11; 82 X.21, 24; 83 X.31
Essay, 5
Eutectic, 49 P.16
Examples, 9
Explosions, 38 K.31; 40 K.39
Extensive property, 67 T.3

Films, insoluble, 56 S.7, 10; 58 S.28
Fluorescence, 61 Sp.12
Force constant, 61 Sp.11
Franck–Condon principle, 61 Sp.11
Frequency factor, 35 K.11

Gas constant, 81 X.15
Gas laws, 34 K.1
Gibbs adsorption equation, 56 S.10; 57 S.21; 58 S.28
Gibbs free energy, 24 E(c).12, 14, 16; 71 T.30, 31, 33, 34, 37; 73 T.50, 55, 58
Gibbs–Duhem equation, 72 T.46
Graphs, 9
Grotthus–Draper law, 62 Sp.17

Half-change time, 35 K.6; 39 K.36
Hammett acidity functions, 30 E(f).5
Heat capacity, 67 T.7–10; 68 T.11
Heisenberg principle, 77 W.8
Henry's law, 47 P.2, 3, 6; 72 T.41
Hess's law, 66 T.1; 67 T.5
Hittorf method, 18 E(a).2
Hydrogen bond, 78 W.14
Hydrolysis, 29 E(e).1; 30 E(f).1

Ideal solutions, 47 P.6, 7; 50 P.24
Indicators, 30 E(f).1, 2, 5; 82 X.25

Intensive property, 67 T.3
Intermolecular forces, 43 M.3
Ion-atmosphere, 84 X.33
Ion-exchange resins, 57 S.20
Ionic mobility, 18 E(a).1, 4, 5; 20 E(b).6; 31 E(f).7; 82 X.20
Ionic product, 29 E(e).2; 31 E(f).9
Ionization potential, 44 M.11
Iso-electric point, 56 S.16

Joule–Thomson effect, 72 T.40, 42; 82 X.22, 23; 83 X.31

Kinetic theory, 34 K.1
Kirchhoff's equations, 74 T.57
Kohrausch's law, 20 E(b).6; 82 X.23

Langmuir's theory, 39 K.35
Latent heat, 60 T.24; 48 P.8; 70 T.26, 27
Lattice energy, 67 T.5
Lindemann theory, 39 K.35, 37
Liquefaction, 81 X.13; 82 X.20

Magnetic moment, 61 Sp.8
Maxwell–Boltzmann law, 73 T.47
Mean free path, 38 K.33
Melting-point, congruent, 49 P.19
Melting-point, incongruent, 49 P.18; 84 X.34
Molecular orbitals, 77 W.5; 78 W.14
Molecular weight, 80 X.1–7; 81 X.8, 9
Molecularity, 34 K.4, 5; 39 K.36, 38
Moment of inertia, 61 Sp.9
Monolayers, 56 S.9
Morse curve, 61 Sp.11

Nernst heat theorem, 74 T.57
Neutralization, heat of, 29 E(e).2
Notes, 5

Order of reaction, 34 K.3–5; 35 K.6–10; 36 K.13, 15; 39 K.36, 38
Osmotic pressure, 81 X.9; 82 X.22

Partial molar property, 72 T.46
Peritectic, 49 P.16

pH, 21 E(b).10; 23 E(c).7; 24 E(c).11; 28 E(d).1-4; 30 E(f).4, 5; 31 E(f).8, 9
Polarizability, 61 Sp.11
Polarization, electrolytic, 17
Polarography, 31 E(f).7; 57 S.24; 82 X.26
Potential
electrode, 22 E(c).1, 2; 23 E(c).7, 9; 24 E(c).11, 13, 17
liquid junction, 16
Potential energy curves, 61 Sp.3
Potentiometric titration, 30 E(f).2; 31 E(f).7, 10

Quantum efficiency, 84 X.33
Quantum numbers, 61 Sp.5, 7
Quantum theory, 77 W.3, 4, 9, 11
Quantum yield, 35 K.12; 62 Sp.17

Raoult's law, 47 P.1-3, 5, 6; 72 T.41, 43
Rate constant, 34 K.3; 35 K.6, 7, 9; 36 K.13, 16, 17; 37 K.22
Refractivity, 82 X.24
Relaxation effect, 16; 20 E(b).5
Retrograde solubility, 84 X.34
Reversible reaction, 35 K.10; 36 K.19
Review, 5
Rotation spectra, 61 Sp.10

Salt bridge, 22 E(c). 3, 4
Salt effects, 81 X.18
Schrödinger's theory, 77 W.4, 6-8, 12
Schulze-Hardy rule, 56 S.16

Solubility, 20 E(b).6, 8; 25 E(c).20; 30 E(f).1, 4
Specific heat, 58 S.27; 83 X.29; 77 W.10; 82 X.22
Spectrophotometry, 31 E(f).8
Stability constant, 29 E(e).6
Steady-state hypothesis, 39 K.4
Surface tension, 57 S.22; 83 X.31

Temperature coefficient, 36 K.15
Third-order reactions, 39 K.34
Transition state, 34 K.4; 35 K.12; 36 K.18; 37 K.25; 39 K.35
Transport number, 18 E(a).1-5; 30 E(f).4; 31 E(f).8, 9
Trouton's rule, 72 T.40
Tyndall effect, 56 S.16

Ultra-filtration, 56 S.16
Unimolecular reaction, 35 K.7; 36 K.15; 38 K.32; 39 K.35

van der Waals's equation, 34 K.1; 81 X.12; 82 X.23
Vapour pressure, 70 T.25-29
Velocity constant: *see* Rate constant
Vibration frequency, 61 Sp.9
Viscosity of gases, 38 K.33
Volume, critical, 30 E(f).4

Walden's rule, 17

X-ray diffraction, 43 M.4, 5, 7; 44 M.12; 77 W.6; 83 X.31

Zeta-potential, 56 S.11; 58 S.29